Structural Geology of the Colorado Plateau Region of Southern Utah, with Special Emphasis on Deformation Bands

George H. Davis
Department of Geosciences
University of Arizona
Tucson, Arizona 85721
USA

THE GEOLOGICAL SOCIETY OF AMERICA
·1888·

SPECIAL PAPER

342

1999

Published by The Geological Society of America, Inc.
3300 Penrose Place, P.O. Box 9140, Boulder, Colorado 80301

Printed in U.S.A.

GSA Books Science Editor Abhijit Basu

Library of Congress Cataloging-in-Publication Data

Davis, George H. (George Herbert), 1942-
Structural geology of the Colorado Plateau region of southern Utah, with special emphasis on deformation bands / George H. Davis.
p. cm. -- (Special Paper ; 342)
Includes bibliographical references.
ISBN 0-8137-2342-6
1. Geology, Structural--Utah. 2. Geology, Structural--Colorado Plateau. I. Title. II. Special papers (Geological Society of America) ; 342.

QE627.5.U8 .D38 1999
551.8'09792--dc21

Cover: Fault fin landscape created by the presence of deformation band shear zones in Navajo Sandstone near the northern end of the Kaibab Uplift, southern Utah. Photograph by George H. Davis.

10 9 8 7 6 5 4 3 2 1

Dedicated to the "Senior Partner"
Peter J. Coney
who simply loved the Colorado Plateau

Contents

Geological Society of America
Special Paper 342
1999

Structural Geology of the Colorado Plateau Region of Southern Utah, with Special Emphasis on Deformation Bands

ABSTRACT

Mesozoic rocks within the Colorado Plateau region of southern Utah bear the structural impact of three major regional tectonic deformations. The first deformation was during the Laramide, from ca. 90 to 50 Ma, when an east-northeast–west-southwest plate-tectonic convergence between the North American plate to the east and the Farallon plate to the west generated horizontal compression that contracted the Colorado Plateau region. Contraction was accommodated by crustal-scale basement-cored thrust, reverse, and oblique-slip faulting and associated folding, expressed in Mesozoic and uppermost Paleozoic stratigraphy by the Kaibab uplift, the Circle Cliffs uplift, and the San Rafael Swell, as well as the great monoclines that flank each uplift. Moreover, the Laramide contraction was accommodated by numerous broad anticlines and synclines, notably within the Kaiparowits Plateau. The second deformation took place during a part of the Miocene, from ca. 25 to 19 Ma, when the rate of plate-tectonic convergence had slackened from ~100 mm/yr during the Laramide to ~50 mm/yr and when relatively shallow subduction was rolling back to a steeper angle. Magmatism induced by subduction was well developed in the foreland region of the Colorado Plateau, and it produced the laccoliths and associated domes of the Henry Mountains, the Marysvale volcanic field and associated laccoliths, and the Paunsaugunt thrust belt. The thin-skinned folding and thrusting that characterizes the Paunsaugunt thrust system deformed rocks as young as the middle Eocene Claron Formation. This thrusting was not a response to regional contraction, but instead was a consequence of emplacement, collapse, and spreading of the southern part of the Marysvale volcanic field at ca. 20 Ma. The third deformation was Basin and Range extensional faulting, which was initiated at ca. 15 Ma and is still active today. Coeval with right-lateral strike-slip movement along the San Andreas transform boundary between the North American plate and the Pacific plate, Basin and Range extension produced the three major high-angle normal faults located within the Western High Plateaus along the southwestern margin of the Colorado Plateau. These are the Hurricane, Sevier, and Paunsaugunt faults.

Important aspects of the kinematics and mechanics of formation of each of the major regional structures are recorded in outcrop and larger-scale shear zones that occur in porous sandstones as deformation bands and zones of deformation bands. They are abundant in the Navajo Sandstone and the Entrada Formation (both are Jurassic) and in certain porous sandstones within the Straight Cliffs Formation (Upper Cretaceous), especially where these formations crop out along major regional structures. Single deformation bands of tectonic origin accommodate millimeter-scale "fault" displacements through a shearing that involves the combination of pore-space collapse, grain-scale fracturing, and cataclasis. This shearing of porous sandstone is a strain-hardening deformation mechanism, which results in an abrupt "lock-up" of

Davis, G. H., 1999, Structural Geology of the Colorado Plateau Region of Southern Utah, with Special Emphasis on Deformation Bands: Boulder, Colorado, Geological Society of America Special Paper 342.

the deformation band. Thus, the first-formed deformation band becomes a stress and strain concentrator for the next increment of deformation-band shearing that will occur alongside. In this way, zones of deformation bands evolve that are centimeters to meters thick and that accommodate centimeters to meters of displacement. Semi-brittle shearing of this type can be superseded by ordinary faulting, which can accommodate meters or 10s of meters of displacement. The faulting is recorded not only by offsets, but also by the presence of slickenlined or grooved slickensided surfaces located on the outer walls of zones of deformation bands. Because of increased mechanical strength achieved through loss of porosity, grain-scale comminution, and precipitation of silica within deformation bands and zones of deformation bands, these structures are quite resistant to weathering and erosion. Thus they commonly project from outcrops as if they were resistant veins or dikes, in some cases generating "fault-fin" landscapes.

Deformation-band shear zones of tectonic origin within the region of study commonly are arranged in "Riedel" systems, within which individual synthetic and antithetic zones occur at angles of ~15° and ~75° to the principal displacement zones. Some Riedel systems are conjugate, with principal displacement zones intersecting at ~60°. The antithetic deformation-band shear zones are especially well developed in transfer zones between overstepping synthetic Riedel shears.

Zones of deformation bands associated with Laramide compression include normal faulting along the upper hinge zones of monoclines, thrust and reverse faults near the lower reaches of monoclinal folds, and strike-slip systems along transpressional zones. Zones of deformation bands associated with Miocene thrusting include conjugate thrust systems. Zones of deformation bands associated with Miocene laccolith emplacement in the Henry Mountains reflect a flexural bending, a flexural slip, and radial and circumferential stretching strain. Zones of deformation bands associated with Basin and Range faulting are closely restricted to the fault zones proper and consist typically of two systems of conjugate-normal faults, one system striking parallel to the major fault zone, the other striking perpendicular.

In addition to the deformation bands and zones of deformation bands that are clearly related to specific discrete regional structures, others occur in one of two systems that are regionally widespread and not associated with specific structures. The first of these occupies the upper 10 to 15 m of the Navajo Sandstone throughout most of the region. It consists of multiple conjugate-normal sets of noncataclastic deformation bands that lack slickenlines, slickensides, and Riedel geometries. They accommodated bulk compaction during burial and loading and formed through granular flow enabled in part by pore fluids. The second of these occurs in the Entrada Formation everywhere within the Colorado Plateau region of southern Utah. Noncataclastic, though marked by grain fracturing, these deformation bands and zones of deformation bands tend to strike parallel to the direction of greatest principal stress in the Laramide. They appear to have accommodated subordinate extensional faulting through normal-slip shearing. Some systems formed in association with and at right angles to the Laramide compression, and others, in response to emplacement of Miocene laccoliths.

The deformation-band shear zones have collectively imparted a macroscopic ductility to major regional lithostratigraphic units, especially the Navajo Sandstone. This capacity contributes to the geometric elegance of the Colorado Plateau regional structures. In locations of strong deformation, individual deformation bands and zones of deformation bands are so plentiful and so closely spaced that they essentially pervade the host rock, especially when viewed at subregional scale. Coordinated shear among deformation-band elements within a system can accommodate whatever regional and local strains must accrue in order to create the smoothest possible fault displacement gradients.

Deformation-band shear zones compartmentalize porous sandstone bodies in ways that influence fluid flow. Because porosity is lost within deformation-band shear zones, these structures seal the boundaries of compartments and can dramatically

influence fluid flow paths through the porous sandstones. Thus the structures are natural laboratories for characterizing reservoir-scale deformation.

Overall, the research demonstrates how a relatively stable structural block like the Colorado Plateau can contain a myriad of structural features that not only record major tectonic events but convey important information about both the events and mechanical response of rocks to far-field tectonic stresses. Also, the research reveals the importance of deformation bands in understanding regional tectonic problems, patterns of fluid flow in the crust, and development of landscape.

INTRODUCTION

Background and objectives

I like to refer informally to the Colorado Plateau region of southern Utah as "the southern Utah parks and monuments region" (Fig. 1). The part I have been studying contains Zion Canyon National Park, Cedar Breaks National Monument, Coral Pink Sand Dunes State Park, Bryce Canyon National Park, Kodachrome Basin State Park, Grand Staircase–Escalante National Monument, Petrified Wood State Park, Glen Canyon National Recreation Area, Capitol Reef National Park, and Goblin Valley State Park (Fig. 1). The structural architecture of this region (Fig. 2) was fashioned by three major tectonic deformations in the Mesozoic and Cenozoic. Laramide deformation (uppermost Cretaceous to middle Eocene) created the Kaibab uplift, the Circle Cliffs uplift, and the San Rafael Swell; the major monoclines associated with these uplifts; and numerous broad, gentle anticlines and synclines (Fig. 2D). Pre–Basin and Range, Miocene deformation generated laccoliths and associated domal flexures in the Henry Mountains (see Fig. 2D) and in the Markagunt Plateau. In addition, Miocene tectonic activity created the Paunsaugunt thrust system (Merle et al., 1993) (Fig. 2C). Basin and Range deformation (late Miocene to present) produced the three major high-angle normal fault systems in the Western High Plateaus of Utah: namely, the Hurricane, Sevier, and Paunsaugunt faults (see Fig. 2C).

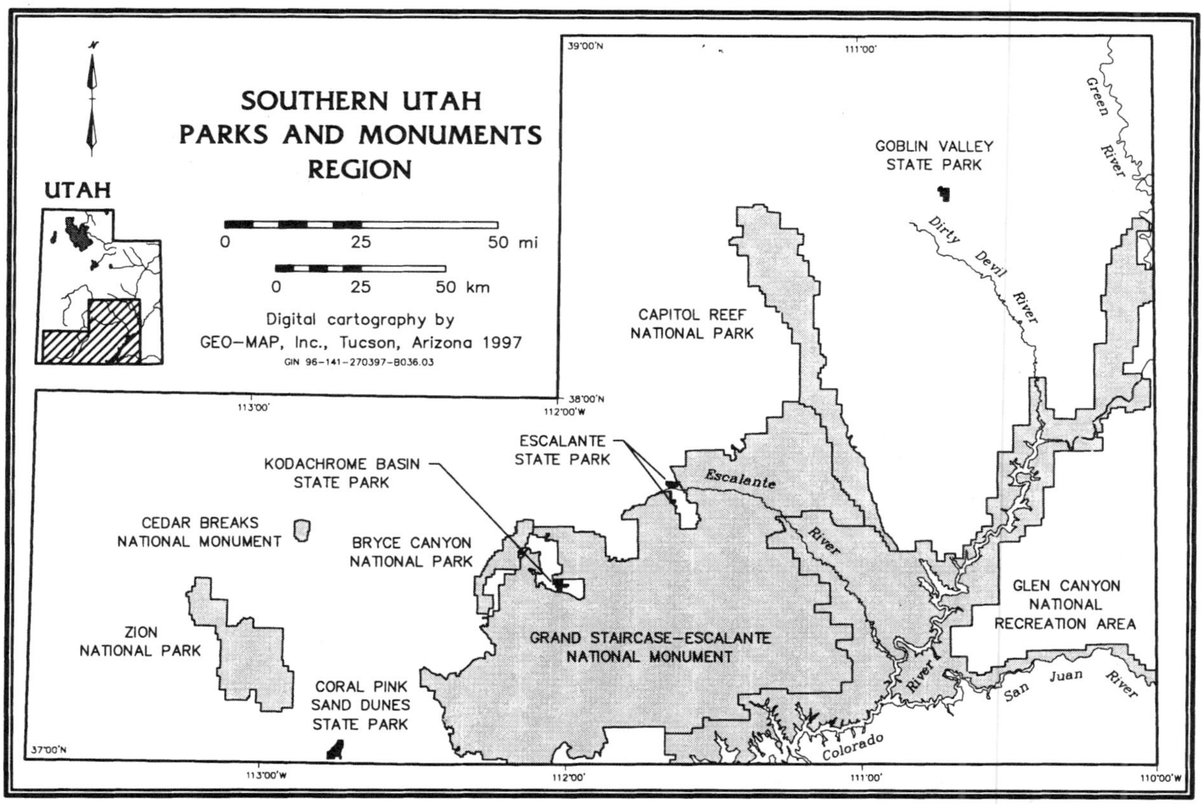

Figure 1. Location of the southern Utah parks and monuments region.

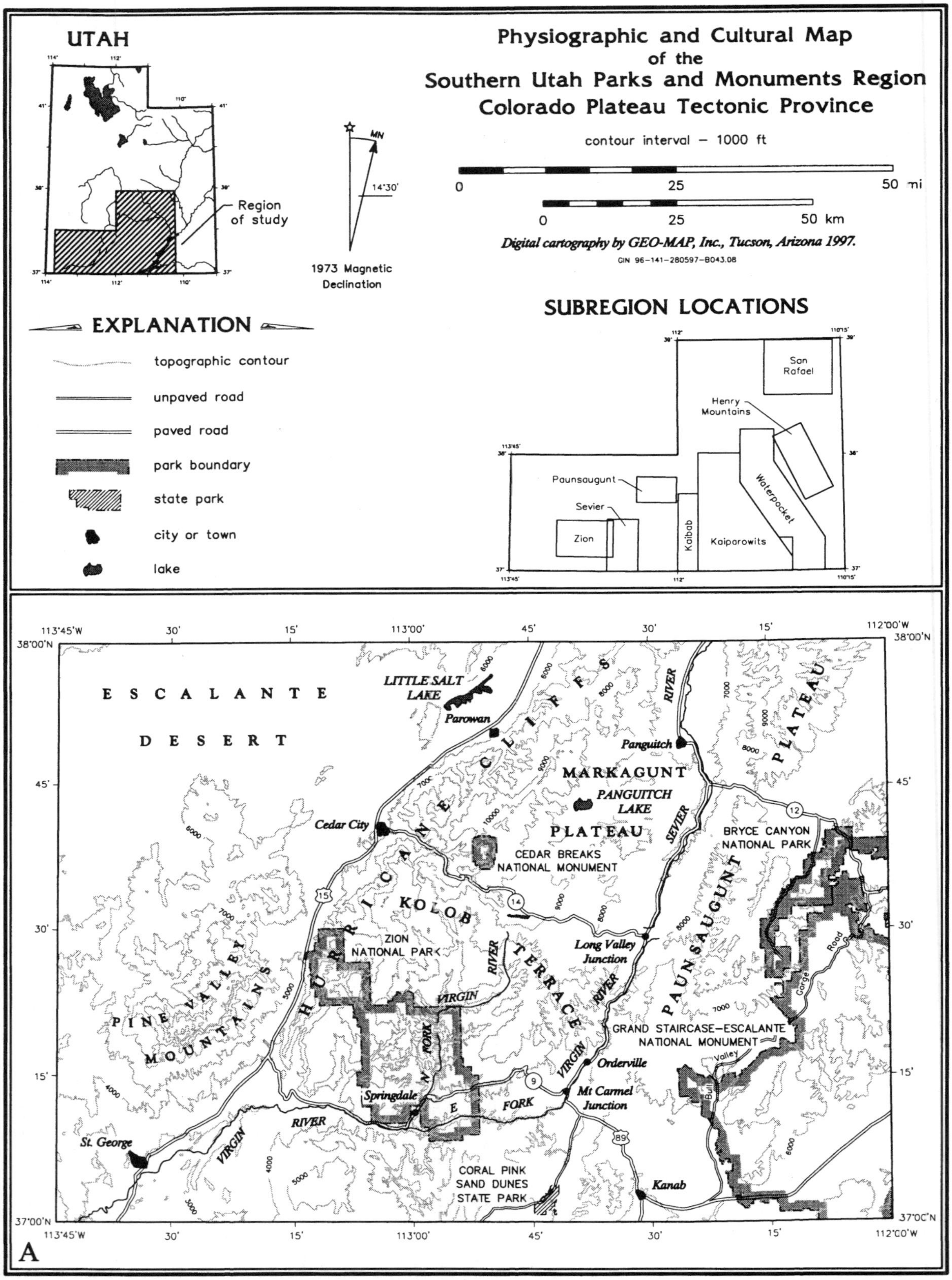

Figure 2 (this and following three pages). A: Physiographic and cultural map of the western part of the region. Inset identifies locations and names of subregions. B: Physiographic and cultural map of the eastern part of the region. C: Structural geology map of the western part of the region. D: Structural geology map of the eastern part of the region. Note that (A) and (B) are topographic maps, and (C) and (D) are structure contour maps.

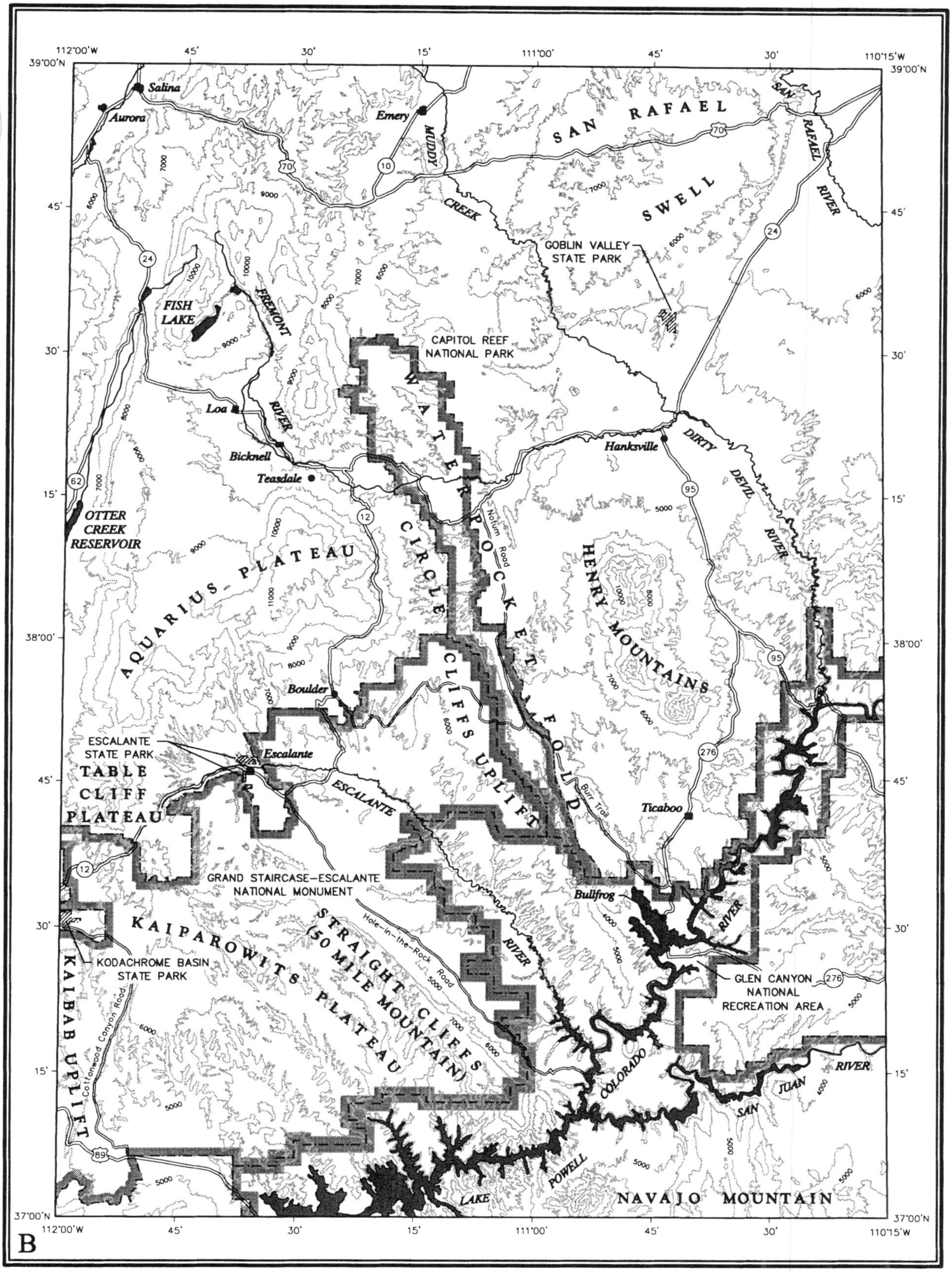
112°00'W
45'
30'
15'
111°00'
45'
30'
110°15'W
39°00'N
37°00'N
38°00'
Salina
Aurora
Emery
MUDDY CREEK
SAN RAFAEL SWELL
SAN RAFAEL RIVER
GOBLIN VALLEY STATE PARK
FISH LAKE
FREMONT RIVER
CAPITOL REEF NATIONAL PARK
Loa
Bicknell
Teasdale
Hanksville
DIRTY DEVIL RIVER
OTTER CREEK RESERVOIR
WATERPOCKET FOLD
CIRCLE CLIFFS UPLIFT
Notom Road
HENRY MOUNTAINS
AQUARIUS PLATEAU
Boulder
ESCALANTE STATE PARK
Escalante
TABLE CLIFF PLATEAU
ESCALANTE RIVER
Burr Trail
Ticaboo
GRAND STAIRCASE–ESCALANTE NATIONAL MONUMENT
Bullfrog
Hole-in-the-Rock Road
KAIPAROWITS PLATEAU
STRAIGHT CLIFFS (50 MILE MOUNTAIN)
KODACHROME BASIN STATE PARK
KAIBAB UPLIFT
Cottonwood Canyon Road
GLEN CANYON NATIONAL RECREATION AREA
COLORADO
SAN JUAN RIVER
LAKE POWELL
NAVAJO MOUNTAIN

B

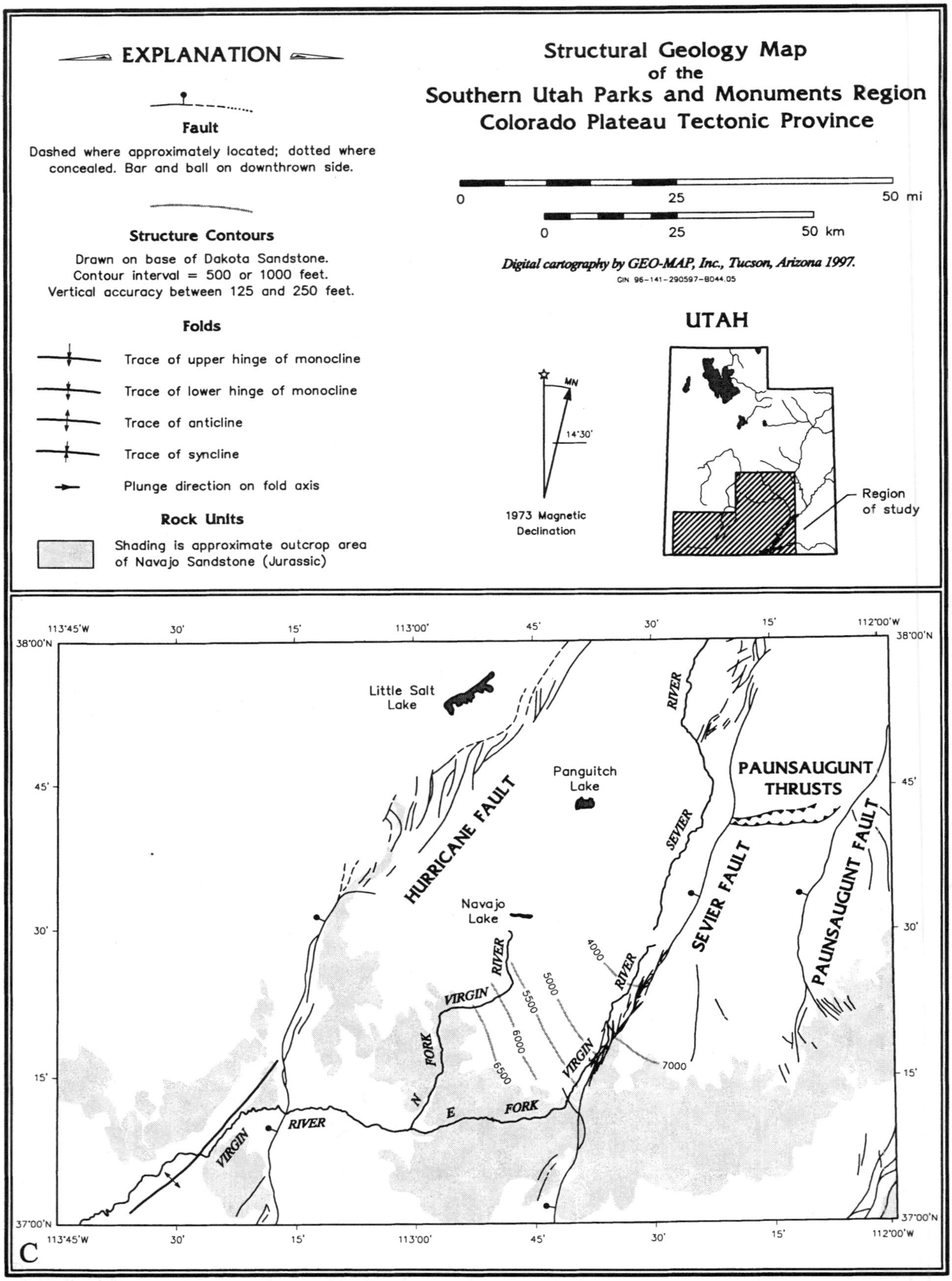

EXPLANATION
Fault
Dashed where approximately located; dotted where concealed. Bar and ball on downthrown side.
Structure Contours
Drawn on base of Dakota Sandstone.
Contour interval = 500 or 1000 feet.
Vertical accuracy between 125 and 250 feet.
Folds
Trace of upper hinge of monocline
Trace of lower hinge of monocline
Trace of anticline
Trace of syncline
Plunge direction on fold axis
Rock Units
Shading is approximate outcrop area of Navajo Sandstone (Jurassic)
Structural Geology Map
of the
Southern Utah Parks and Monuments Region
Colorado Plateau Tectonic Province
0
25
50 mi
0
25
50 km
Digital cartography by GEO-MAP, Inc., Tucson, Arizona 1997.
GIN 96-141-290597-B044.05
UTAH
MN
14°30'
1973 Magnetic Declination
Region of study
113°45'W
30'
15'
113°00'
45'
30'
15'
112°00'W
38°00'N
45'
30'
15'
37°00'N
Little Salt Lake
Panguitch Lake
Navajo Lake
HURRICANE FAULT
PAUNSAUGUNT THRUSTS
SEVIER FAULT
PAUNSAUGUNT FAULT
SEVIER
RIVER
VIRGIN
RIVER
N FORK
E FORK
4000
5000
5500
6000
6500
7000
C

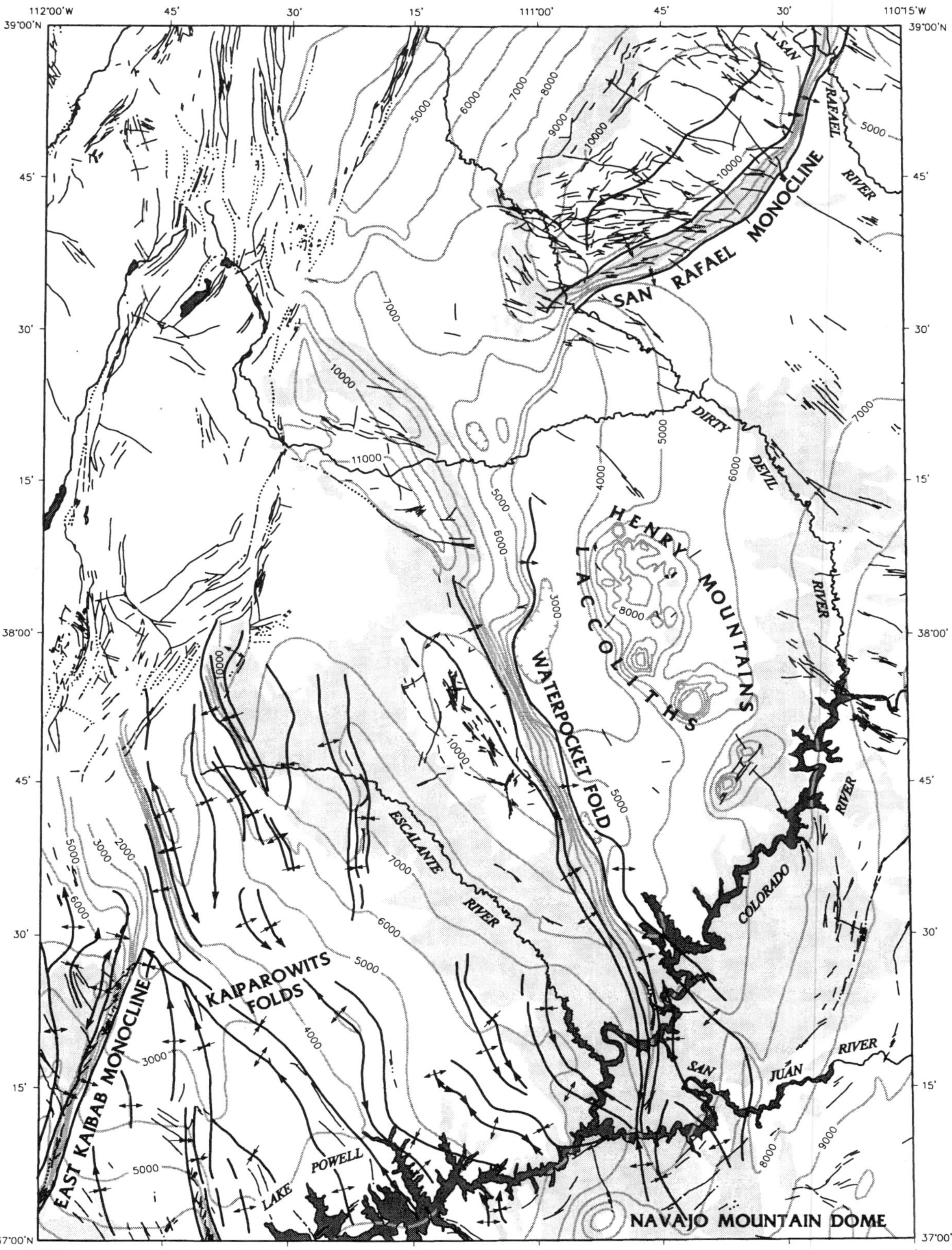

SAN RAFAEL MONOCLINE
SAN RAFAEL RIVER
DIRTY DEVIL RIVER
HENRY MOUNTAINS LACCOLITHS
WATERPOCKET FOLD
ESCALANTE RIVER
COLORADO RIVER
SAN JUAN RIVER
KAIPAROWITS FOLDS
EAST KAIBAB MONOCLINE
LAKE POWELL
NAVAJO MOUNTAIN DOME
112°00'W
110°15'W
39°00'N
38°00'
37°00'N
111°00'

D

Where high-porosity quartz sandstones, such as those of the Navajo Sandstone (Jurassic), have been deformed by the monoclinal folding, laccolithic doming, thin-skinned thrusting, and high-angle normal faulting, the sandstones are pervaded by deformation bands (Fig. 3, A–C). Deformation bands are thin, semibrittle shear zones along which small fault offsets have been accomplished (Aydin, 1978). At inception, they average only several millimeters in thickness and may remain so. But commonly as one deformation-band shear zone work hardens, a new one forms alongside; thus zones of deformation bands typically develop, ranging in thickness from 1 cm to several meters (Aydin and Johnson, 1978). The outer surfaces of thick zones of deformation bands may be discrete faults, marked by slickensides and slickenlines (Fig. 3D). Such faults might accommodate 10s of meters of displacement.

Through broad-based field research and reconnaissance that I conducted on the Colorado Plateau in the late 1980s and early 1990s, I became aware of the abundance and wide distribution of deformation-band shear zones in the southern Utah parks and monuments region. Because of this finding, I decided to carry out detailed structural analysis of the geometry and kinematics of deformation-band shear zones within a 40,000 km^2 region of south-central Utah (Figs. 1 and 2). The purpose was both structural and regional tectonic in nature. First, I wanted to explore the occurrences and characteristics of deformation bands and zones of deformation bands in several structural and tectonic settings and in a variety of formations, in order to gain insight regarding their origins. Indeed, the literature documents quite a number of deformation-band fabrics and structures, making it essential to discriminate among the types of deformation bands. Second, I wished to use the orientations and kinematic characteristics of the sets of deformation bands to evaluate strain patterns and, where possible, stress patterns related to the development of specific regional structures. For instance, deformation bands have enormous potential for reconstructing the progressive deformation of a given structure. They form very readily in response to stresses in porous sandstones, and as a result they occur in great numbers along major structures. Individual millimeter-scale deformation bands "strain harden" almost as soon as they form. Strain hardening takes place during grain crushing and cataclastic flow. Friction within a deformation band increases to the point that further shear is impossible. Consequently, subsequent increments of deformation are achieved by repetition of the process in immediately adjacent porous sandstone (Aydin and Johnson, 1978). Thus, when individual deformation bands lock up, they are preserved collectively as increments of the strain history. Furthermore, deformation bands do not commonly reactivate. Therefore, the confusion embodied in analysis of kinematics of systems of reactivated structures, such as joints and faults, is avoided.

As a third objective, I wanted to see if I could recognize broad regional patterns of strain and stress representing the Laramide, Miocene (ca. 25–20 Ma), and Basin and Range deformations. Conceptually, this third research objective included exploring for tectonic strain patterns expressed in deformation bands within subregions of horizontal strata outside of the locations of major regional structures, such as monoclines or long continuous fault zones. The motivation for this objective stemmed from a reconnaissance-based knowledge of the abundance of deformation bands in perfectly flat lying, porous Navajo Sandstone in places such as the highest mesas and plateaus of Zion National Park, in locations quite removed from major structures. If the regional strain and stress patterns could be mapped in such areas, insight could be achieved regarding the manner in which tectonic stresses are transmitted great distances from plate boundaries into foreland and continental-interior regions.

A fourth research objective was identified after the start of the project, namely, to evaluate the nature of the "compartmentalization" of porous sandstone by sets and systems of deformation-band shear zones. The mechanism of formation of deformation-band shear zones is such that porosity is almost completely eliminated within these zones. As a result, deformation-band shear zones are commonly barriers to fluid flow and thus may profoundly influence permeability flux and tortuosity within a porous sandstone host (Antonellini and Aydin, 1994, 1995). I investigated two areas in particular with the goal of closely mapping the orientation and spacing of sets of deformation-band shear zones and interpreting the patterns in relation to compartmentalization.

Methodological approach

This work represents primarily four field seasons of data collection and structural analysis (1994–97), although it also builds upon mapping of Colorado Plateau monoclines (Davis and Kiven, 1975; Davis, 1978); discovery and evaluation of Miocene thrusts (Davis and Krantz, 1986; Lundin and Davis, 1987; Davis, 1990; Nickelsen et al., 1992; Merle et al., 1993); and discovery and mapping of transpressive, reverse, right-lateral faulting along the East Kaibab monocline (Davis and Tindall, 1996; Tindall and Davis, 1999). One result of these field studies was a close familiarity with the properties of the major regional structures in the Colorado Plateau.

With regard to the focused work on deformation bands, my field work in 1994 and 1995 was directed primarily toward gathering orientations and evaluating kinematic properties of deformation bands and zones of deformation bands across the entire southern Utah parks and monuments region (Figs. 1, 2). My field work in 1996 and 1997 was focused primarily on detailed mapping and analysis of especially well exposed systems of deformation-band shear zones. Several rock formations proved to be the most productive hosts for deformation bands and deformation-band shear zones: Navajo Sandstone (Jurassic), the Page Sandstone Tongue of the Carmel Formation (Jurassic), Entrada Sandstone (Jurassic), and Straight Cliffs Sandstone (Upper Cretaceous). My students and I (see Acknowledgments) measured the orientations and kinematic properties of ~3,000 deformation bands and deformation-band shear zones at 78 stations during the course of the regional inventory. In this paper, the 78 stations are grouped within eight subregions. In addition, my students and I

Figure 3. A: Hand specimen of deformation bands in Navajo Sandstone (Jurassic), Sheets Gulch area, Waterpocket Fold. Note tiny offset along individual deformation bands. B: Photomicrograph of this same specimen, showing deformation band cutting through porous host. Quartz within the band is highly crushed. Host rock is marked by larger grains and plenty of pore space (dark intergranular areas is epoxy filling). Width of field of view is 320 μm. C: Photograph of thick deformation-band shear zone in Navajo Sandstone (Jurassic), Cottonwood area, East Kaibab monocline. Note camera bag for scale. D: Photograph of surface of one of the outer walls of a deformation-band shear zone, showing slickensides and slickenlines. Protolith is Navajo Sandstone. Location is Capitol Reef National Park in Capitol Gorge.

measured the orientations and kinematic properties of ~800 deformation bands and deformation-band shear zones during detailed structural mapping (1:600) of two outstanding exposures of deformation bands: the Cottonwood area at the north end of the Kaibab uplift and the Sheets Gulch area at the northern end of the Waterpocket Fold in Capitol Reef National Park.

For each deformation band or deformation-band shear zone, I tried to secure the following data: strike, dip, nature of offset, magnitude of offset, slip direction, and thickness. Not all outcrops yielded offset and slip-direction information. I took ~1,500 photographs of deformation bands and zones of deformation bands and the host-rock units and structures. Rock samples were collected, from which thin sections (doubly polished) were prepared for examination of microstructures. Samples were impregnated with blue epoxy, thus permitting easy recognition of porosity in thin section. Photographic enlargements of outcrop relationships were made, from which tracings were drawn of the geometries. Photomicrographs of thin sections were produced to illustrate the internal properties of deformation-band shear zones.

A FoxPro database management system was programmed (see Acknowledgments) to organize and operate on the database. Primary output was in the form of stereographic projections of raw orientation data and rose diagrams of trends. Stereographic projections were made using Richard W. Allmendinger's "STEREONET v.4.9.5, Academic Version," which he kindly made available to colleagues. Furthermore, geography, structure contours, and the outcrop expression of the Navajo Sandstone were digitized for the region of study and entered into an AutoCad system for preparing base maps of subregions. The Cottonwood and Sheets Gulch areas, each of which is ~1 km^2 in area, were given special attention because of the extraordinary quality of the topographic expression of their deformation-band shear zones. In preparation for aerial-photographic fly-overs, 10 survey targets were placed in each of the areas, and their elevations were determined by using GPS (Global Positioning System) surveying instrumentation for control. Low-altitude photographic coverage was then flown and used stereoscopically and cartographically to prepare 1:600 topographic base maps, with 3 m (10 ft) contour intervals. Enlarged aerial photographs were best-fit rectified and produced at the same matching 1:600 scale (see Acknowledgments).

By using photographs, sketches, thin sections, field notes, and the data sets, I classified the deformation bands and zones of deformation bands according to types (see Appendix Table 1); and for each type and location, I interpreted the kinematics. I identified by subregion the orientations of deformation bands and zones of deformation bands of known kinematics (e.g., thrust, strike-slip, normal, and/or oblique slip). Then, I evaluated the degree to which the kinematic characterization could be extended to other deformation bands and zones of deformation bands on the basis of identity of orientation and physical character.

Organization of text

This Geological Society of America Special Paper consists of two main parts. Part I, Major Regional Structures, begins with a description of the geography of the southern Utah parks and monuments region. This is followed by a brief description of the regional stratigraphy. Part I is devoted to description of the major regional geologic structures, beginning with Basin and Range faults, namely the Hurricane, Sevier, and Paunsaugunt faults; then focuses on pre–Basin and Range (Miocene) structural systems, namely, the Paunsaugunt thrust system and the Henry Mountains laccoliths; and next presents Laramide structures, including the major Colorado Plateau basement-cored uplifts, associated monoclines, and systems of anticlines and synclines. Part I is intended to provide context for the significantly longer Part II, Deformation-Band Shear Zones, which is a detailed overview of the nature and occurrences of deformation-band shear zones across the southern Utah parks and monuments region.

Part II begins with a review of the literature on deformation bands and zones of deformation bands. The review is then followed by a subregion by subregion description of outcrop properties of the deformation-band shear zones; a presentation of the stereographic geometry of deformation-band shear zones that were measured and/or mapped; and an interpretation of origin of deformation-band shear zones in relation to the formation of major regional structures.

In the Conclusions, the formation of the several varieties of deformation-band shear zones is discussed in both a structural geologic and a regional tectonic context. Furthermore, within the Conclusions, I comment on the significance of reservoir-rock compartmentalization and call attention to the influence of deformation-band shear zones on the formation of landscapes.

Part I: Major regional structures

GEOGRAPHY

Physiographic subprovinces

The Colorado Plateau of the western United States is renowned for its extraordinary landscapes and geography. Truly it is one of the most sublime geologic provinces in the world. The attraction of the landscape of the Colorado Plateau resides in several factors, all of which are illustrated by the southern Utah parks and monuments region: expansive vistas of cliffs, buttes, mesas, and plateaus; deeply incised canyons, including its narrow "slot" canyons; colorful, stark, textured rock formations; forested high plateaus; the mix of water and desert; and the endless backcountry. Geology and tectonic history of the Colorado Plateau are responsible for this landscape, one so unique that the whole region seems to have been set aside for parks and monuments.

The physiography and structure of the southern Utah parks and monuments region are presented in Figure 2. Three distinctive geographic subprovinces can be identified: the Western High Plateaus on the west (Fig. 2A), which consist of the Kolob Terrace, the Markagunt Plateau, and the Paunsaugunt Plateau; the Henry Mountains on the east (Fig. 2B), which consist of five distinctively domical mountains; and the Laramide uplifts region in the middle (Fig. 2B), which includes the Kaibab uplift, the Circle Cliffs uplift, the Miners Mountain uplift, and the San Rafael Swell.

The Western High Plateaus each are demarcated along their western and eastern borders by high-angle Basin and Range faults, namely the Hurricane, Sevier, and Paunsaugunt faults (Fig. 2C). The Kolob Terrace and Markagunt Plateau (Fig. 2A) are bordered on the west by the Hurricane fault and on the east by the Sevier fault (Fig. 2C). The Paunsaugunt Plateau (Fig. 2A) is bordered on the west by the Sevier fault and on the east by the Paunsaugunt fault (Fig. 2C). Beautiful exposures of the Claron Formation (Eocene) give this subregion its scenic distinctiveness, for the Paunsaugunt Plateau itself is bounded on the west, south, and east by the Pink Cliffs that consist of Claron strata. The eastern escarpment of the Paunsaugunt Plateau is Bryce Canyon National Park. The elevation of the Paunsaugunt Plateau is ~2,400 m (7,900 ft), and fully 450 m (1,485 ft) of topographic relief separate the average elevation of the Paunsaugunt Plateau from the Tropic basin directly to the east. The northern limit of the Paunsaugunt Plateau is ill defined, for the Pink Cliffs gradually give out (Fig. 2, A and C), and the Claron Formation becomes covered by Miocene volcanic rocks.

The Kolob, Markagunt, Paunsaugunt, and Table Cliff plateaus together compose the Grand Staircase (Fig. 4), distinguished by ascending cliffs of the most resistant formations within the stratigraphic sequence: Chocolate Cliffs (Moenkopi Formation, Triassic), Vermillion Cliffs (Kayenta Formation, Jurassic), White Cliffs (Navajo Sandstone, Jurassic), Gray Cliffs (Straight Cliffs Formation, Cretaceous), and Pink Cliffs (Claron Formation, Eocene). Within the Kolob Terrace (Fig. 2A), the expression of the Grand Staircase begins at an elevation of ~1,200 m (3,960 ft) at the base of the Navajo Sandstone at Zion National Park and climbs to an elevation of ~3,200 m (10,560 ft) at the top of the Claron Formation at Cedar Breaks National Monument. Within the Paunsaugunt Plateau (Fig. 2A), the expression of the Grand Staircase begins at an elevation of ~1,500 m (5,000 ft) at the base of the Navajo Sandstone near Coral Pink Sand Dunes State Park and climbs to an elevation of ~2,400 m (7,920 ft) at the top of the Claron Formation in Bryce Canyon National Park. East of the Paunsaugunt fault, in the Table Cliff Plateau region, the top of the cliffs of Claron Formation lies at an elevation of 3,100 m (10,230 ft). Seen in detail, the geographic interruptions of the cliffs, plateaus, and assorted structural terraces are controlled by the traces of the Hurricane, Sevier, and Paunsaugunt fault zones (Fig. 2, A and C).

Within the central and northeastern parts of the southern Utah parks and monuments region, the geography and landscape are fundamentally controlled by the structural geometry of the Laramide uplifts (Kaibab, Circle Cliffs, San Rafael) and associated folds, especially monoclines (Fig. 2, B and D). The expressions of the structures are clearly registered in topography and physiography; the topographic culminations reside at elevations of 1,800 m (5,940 ft), 1,750 m (5,775 ft), and 2,000 m (6,600 ft), respectively. To the east of the Kaibab uplift is a large Creta-

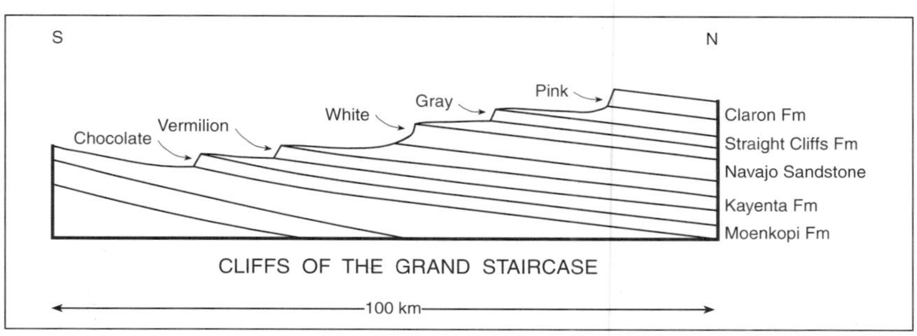

Figure 4. Cliffs of the Grand Staircase, southern Utah. See text for stratigraphic controls.

ceous basin, the Kaiparowits basin, which is expressed as a dissected plateau bordered physiographically on the east by the Straight Cliffs. The top of the Kaiparowits Plateau rests at ~2,000 m (6,600 ft). To the east of the Circle Cliffs uplift and the San Rafael Swell are enormous tracts of lower-elevation (~1,400 m [4,620 ft]) desert landscape—the Cane Spring and the San Rafael Deserts.

The Henry Mountains, which occupy the extreme east-central part of the region of study (Fig. 2, B and D), comprise five major igneous centers (Mount Ellsworth, Mount Holmes, Mount Hillers, Mount Pennell, and Mount Ellen), each marked by a laccolith that domed Mesozoic strata in the Miocene. The centers are highly resistant to erosion, creating individual discrete domical mountains. Mount Ellsworth and Mount Holmes each reside at ~2,500 m (8,250 ft) in elevation. The elevations of Mount Hillers, Mount Pennell, and Mount Ellen are 3,250 m (10,725 ft), 3,450 m (11,385 ft), and 3,500 m (11,550 ft), respectively. To the south-southwest of the Henry Mountains is Navajo Mountain, a dome of Navajo Sandstone underlain by a laccolithic intrusion. Navajo Mountain is 3,148 m (10,388 ft) in elevation (Fig. 2, B and D).

Thus the various geographic subprovinces of the southern Utah parks and monuments region have their own distinctive structure and bedrock control. The Western High Plateaus contain the clearest record of Basin and Range faulting. The Laramide uplifts subprovince illustrates the expressions of Late Cretaceous to middle Eocene contractional deformation. The Henry Mountains subprovince contains the record of Miocene laccolith emplacement.

River systems

There are a number of significant drainage systems in the region. Long Valley Junction is the drainage divide between the Sevier River, which flows northward, and the East Fork of the Virgin River, which flows southward (Fig. 2A). Both of these rivers flow along the general trace of the Sevier fault. At Mount Carmel Junction, the East Fork of the Virgin River abruptly turns westward, where it flows toward St. George. Just south of Springdale (Fig. 2A), the North Fork of the Virgin River, which flows through Zion Canyon, joins the East Fork.

Although not quite distinctive enough to be shown at the scale of the topographic reference map (Fig. 2B), the headwaters of the Paria River run northward close to the general trace of Cottonwood Road, which is the secondary (dirt) road that branches from Highway 89 to the west of Lake Powell and heads northward to Kodachrome Basin State Park. The headwater tributaries of the Paria drainage system are eroding the cliffs that border the eastern margin of the Paunsaugunt Plateau.

Two major rivers flow southeast into Lake Powell (Fig. 2B), namely the Escalante River and the Dirty Devil River. The Escalante River flows along the western margin of the Circle Cliffs uplift, though to the north it bends westward where it crosses, then follows, the Escalante monocline. The Dirty Devil River is the confluence of Muddy Creek and the Fremont River (Fig. 2B). The eastward-flowing Fremont River crosses the Waterpocket Fold at nearly right angles.

Road access

Paved road access into the interior of the southern Utah parks and monuments region is not overdone. Highway 9 passes along the southern margin of the Kolob Terrace and Zion National Park (Fig. 2A). To the north, Highway 14 crosses the Markagunt Plateau. Highway 14 eastward from Cedar City affords a spectacular ascent of the Hurricane Cliffs and passes immediately south of Cedar Breaks National Monument. The steady gentle descent of Highway 14 from Cedar Breaks to Long Valley Junction expresses the modest eastward dip of the entire Markagunt Plateau.

The Arizona-Utah state border area between Lake Powell and Kanab, Utah (Fig. 2, A and B), is accessible via Highway 89, which to the north of Kanab cuts northwestward to Mount Carmel Junction and then northward along Long Valley through Orderville, Long Valley Junction, Panguitch, and beyond. Highway 89 crosses the Kaibab uplift (Fig. 2B); road cuts are spectacular. Scenic Highway 12 branches eastward from Highway 89 south of Panguitch (Fig. 2A), where the road climbs through Red Canyon and then crosses the Paunsaugunt Plateau. This is the highway that leads to the entrance to Bryce Canyon National Park. The main paved road in Bryce Canyon National Park provides access to Rainbow Point, which lies on one of the southern promontories of the Paunsaugunt Plateau.

To the east of Bryce Canyon National Park, Highway 12 descends the eastern margin of the Paunsaugunt Plateau and, in so doing, cuts through a part of the Paunsaugunt thrust system as well as the Paunsaugunt normal-slip fault. As Highway 12 continues eastward to Escalante (Fig. 2B), it passes along the northern margin of the Kaiparowits Plateau. Laramide anticlines and synclines are well exposed in cross-sectional view along this stretch of highway. East of Escalante, Highway 12 turns northward through Boulder to Teasdale (Fig. 2B). The Burr Trail can be followed eastward out of Boulder, where it soon crosses the Circle Cliffs uplift, remaining paved to the very top of the Waterpocket Fold. The Escalante to Teasdale stretch of Highway 12 is one of the most striking in the region, starting out in slickrock exposures of Navajo Sandstone that stretch to the skyline, passing along a narrow rib of rock bordered by deep canyons on either side (Calf Creek on the west, Boulder Creek on the east), and then, north of Boulder, ascending into forested high-mountain topography of Boulder Mountain near the summit of the Aquarius Plateau.

East of Teasdale, Highway 24 passes the entrance to Capitol Reef National Park, cuts through the Navajo Sandstone, and heads off to Hanksville and then northeastward toward Green River, where it connects with Interstate 70. The highway stretch from Hanksville to Green River runs for the most part along the southeastern margin of the San Rafael Swell (Fig. 2B). Interstate 70 crosses the San Rafael Swell along an east-west alignment. To the south of Hanksville, Highway 95 passes along the eastern margin of the northernmost three lobes of the Henry Mountains, and then Highway 276 cuts south-southwestward across a low pass in the Henry Mountains. Highway 276 intersects Lake Powell at Bullfrog Marina. There, the crossing is by ferry, which connects travelers with the continuation of Highway 276 south of the lake.

Going deeper into the interior of the southern Utah parks and monuments region requires backcountry driving on dirt roads. The Bull Valley Gorge Road takes off northward from Highway 89 just east of Kanab (Fig. 2A) and heads along the southeastern margin of the Paunsaugunt Plateau, crossing the Paunsaugunt normal-slip fault. The Bull Valley Gorge Road joins Highway 12 near Cannonville. The Cottonwood Canyon Road splits northward from Highway 89 to the west of Lake Powell (Fig. 2B) and then heads north-northeast to Kodachome Basin State Park. The Cottonwood Canyon Road runs right along the East Kaibab monocline.

The Hole-in-the-Rock Road takes off to the south from Highway 12, just east of Escalante (Fig. 2B), where it can be followed to the cliffs directly above Lake Powell. The Hole-in-the-Rock Road lies just east of the base of the Straight Cliffs (and their continuation into Fifty-Mile Mountain), a spectacular escarpment that marks the eastern edge of the Kaiparowits Plateau. Several rough ranch roads climb westward up into the "Straights" from the Hole-in-the-Rock Road. On the top of the Kaiparowits Plateau there are 10s of winding dirt roads that reflect a history of cattle raising and mining and petroleum exploration. The Hells Backbone Road (not shown) takes off to the north from Escalante, climbing toward the Aquarius Plateau and then shifting eastward toward Boulder through dramatically incised canyon country.

Access along the Waterpocket Fold is provided by the Notum Road, which heads southward from a point just east of the Visitors Center at Capitol Reef National Park. Starting out paved, the road turns to dirt and remains so to where it intersects Burr Trail, which continues to Bullfrog Marina (Fig. 2B). A spectacular switchback road splits off from the Notum Road westward to the top of Burr Trail. This Burr Trail ascent goes right up the Waterpocket Fold at the stratigraphic level of the Navajo Sandstone. Several dirt roads provide access to Mount Hillers, the southernmost of the three Henry Mountains lobes that lie north of Highway 276 (Fig. 2B). Quite a number of dirt roads provide access into the San Rafael Swell. Many of these roads were made during the 1950s when uranium exploration was booming.

REGIONAL STRATIGRAPHY

The map pattern of formations

The geologic formations in the Colorado Plateau region of southern Utah (Fig. 5) are arranged in a simple "layer-cake" fashion, although major regional structures importantly control the structural orientations and the map distributions of units. The full map distribution of the geologic formations in the southern Utah parks and monuments region is displayed in Hintze's (1980) geologic map of the State of Utah. Furthermore, Hintze (1988) has described each of the rock formations in detail, including regional variations in lithologies and depositional environments. I have not attempted here to replicate Hintze's (1980) geologic map. Instead, I have simply shown in Figure 2, C and D, the outcrop trace of the Jurassic Navajo Sandstone in order to provide a stratigraphic reference for describing the map pattern of formations. Indeed, the Navajo Sandstone is a useful point of reference, for it stands out so conspicuously in the landscape of southern Utah by virtue of its thick, resistant, massively cross-bedded nature. Furthermore, it is the dominant host rock to the deformation bands and zones of deformation bands described in the next chapter. Marzolf (1983, p. 636) described the Navajo Sandstone petrographically as "a feldspathic quartz arenite to quartz-rich subarkose composed of rounded to well-rounded quartz sand and containing large to very large-scale trough, tabular-planar, and wedge-planar cross stratification." Within the region of study, the Navajo Sandstone ranges in thickness from ~600 m (1,980 ft) on the west in the vicinity of Zion National Park to ~300 m (990 ft) on the east in the vicinity of Capitol Reef National Park (Marzolf, 1983). During the time of deposition of the Navajo Sandstone, wind direction changed from out of the northwest to out of the northeast (Marzolf, 1983).

In the southwestern part of the region (Fig. 2C), the Navajo Sandstone dips very gently to the north or northeast, and its outcrop trace coincides with the White Cliffs of the Grand Staircase. The narrow east-trending strip south of the White Cliffs and north of the Arizona-Utah state line (Fig. 2C) is occupied by exposures of very gently dipping Permian, Triassic, and Lower Jurassic formations. Stratigraphic thicknesses have been estimated by Cook (1960), Hintze (1980, 1988), Hintze et al. (1994), and Hurlow (1998). The stratigraphy includes (Fig. 5A) Kaibab Limestone (~250 m [825 ft]; Permian), Moenkopi Formation (~520 m [1,716 ft]; Triassic), Chinle Formation (~160 m [528 ft]; Triassic), Moenave Formation or Wingate Sandstone (~130 m [429 ft]; Triassic), and Kayenta Formation (~190 m [627 ft]; Triassic). The Navajo Sandstone, which defines the outcrop expression of the White Cliffs, is Jurassic in age and is ~700 m (2,310 ft) thick. Jurassic to Eocene strata lie above the Navajo Sandstone of the White Cliffs (Figs. 2C and 5A), namely, Carmel Formation (~210 m [693 ft]; Jurassic), Entrada Formation (~130 m [420 ft]; Jurassic), Straight Cliffs Formation (~320 m [1,056 ft]; Cretaceous), and Claron Formation (~210 m [693 ft]; Eocene) (Hintze, 1980, 1988). All of these strata of the Kolob Terrace and Markagunt Plateaus compose a very gently dipping homoclinal sequence.

The Basin and Range Sevier fault disrupts the east-west continuity of the Navajo Sandstone as well as all other Triassic to Eocene formations (Fig. 2C), for the Sevier fault has accommodated ~750 m (2,475 ft) of down-to-the-west, normal displacement. Between the Sevier and Paunsaugunt faults and from the Arizona-Utah state line northward, the Grand Staircase of stratigraphic formations is beautifully laid out (Fig. 5A): red siltstones, sandstones, and mudstones of the Moenkopi Formation (Triassic); gray, pink, and purple claystones, mudstones, and sandstones of

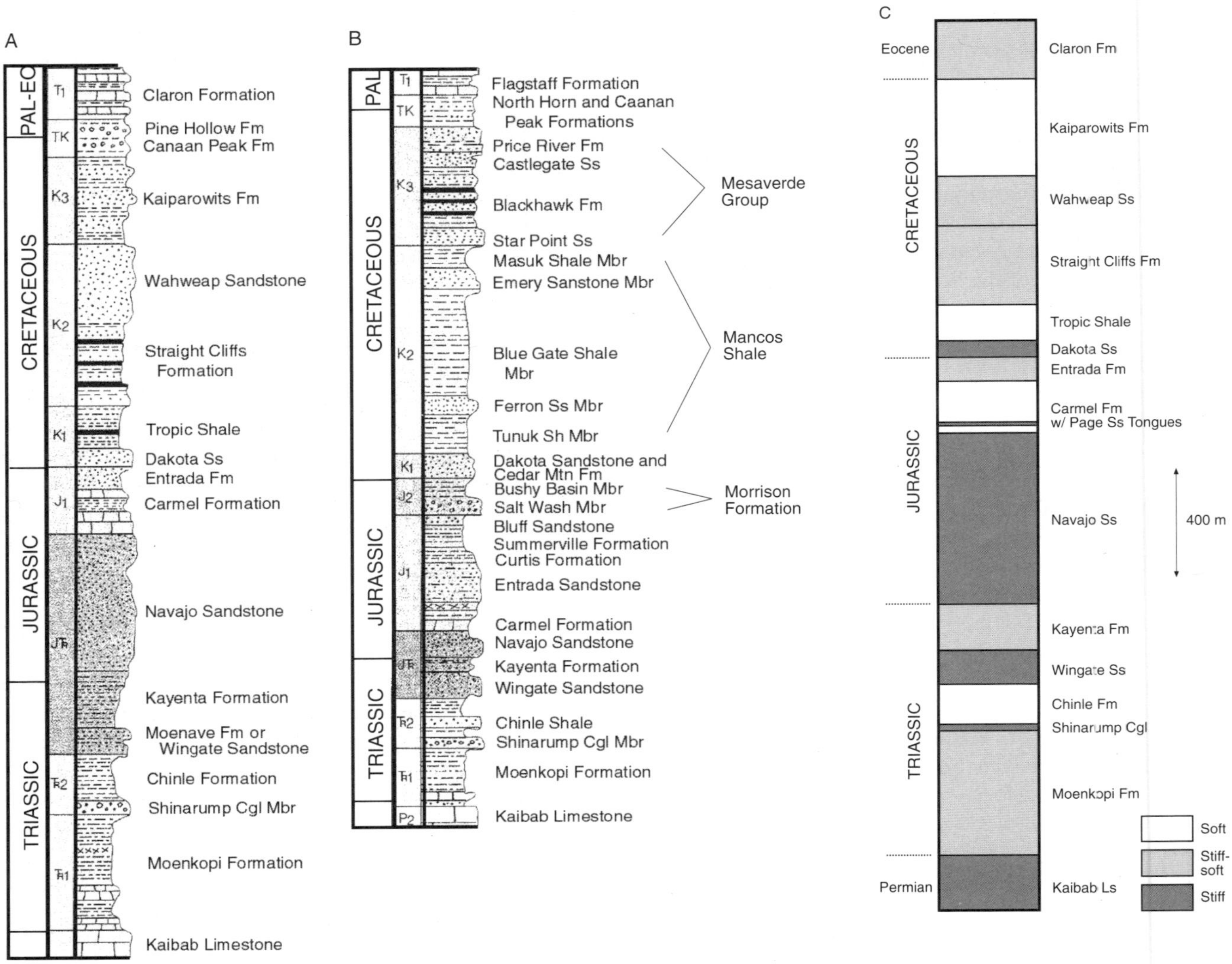

Figure 5. Stratigraphy of the region. A: Stratigraphic column of the western part. B: Stratigraphic column of the eastern part. Both from Hintze (1980). C: Representation of the mechanical stratigraphy of formations within the region.

the Chinle Formation (Triassic), and tan pebble conglomerates of the Shinarump Member at its base (Triassic); white to tan cross-bedded sandstones of the Moenave Formation or Wingate Sandstone (Jurassic); red thick-bedded sandstones, siltstones, and interbedded shales of the Kayenta Formation (Jurassic); white massive cross-bedded sandstones of the Navajo Sandstone (Jurassic); pink siltstones and mudstones, white cross-bedded sandstone tongues, and white or gray anhydrite and gypsum beds of the Carmel Formation (Jurassic); gray shales and black coal of the Tropic Shale (~140 m [462 ft]; Upper Cretaceous); gray to tan sandstones and interbedded gray shales of the Straight Cliffs Formation (Upper Cretaceous); and pink sandstones, siltstones, and conglomerates overlain by white limestones and siltstones of the Claron Formation (Eocene) (Hintze, 1980, 1988) (Fig. 5A).

The Basin and Range Paunsaugunt fault separates the stratigraphy and structure of the Western High Plateaus to the west from the stratigraphy and structure of the Laramide uplifts region to the east. Canyons that have eroded the Kaibab uplift (Fig. 2D) have exposed a continuous, concordant stratigraphy ranging from Permian to Upper Cretaceous. The Permian through Triassic part of this "Kaibab uplift" stratigraphic sequence is essentially the same as that of the Grand Staircase, but important differences show up in the Jurassic and Cretaceous stratigraphy (Fig. 5A). For example, the Carmel Formation contains two white, cross-bedded sandstone units, known as the Page Sandstone Tongue. Each unit of the Page Sandstone Tongue is ~25 m (82 ft) thick. Above the Carmel Formation there is a sequence of red sandstones, siltstones, and interbedded gray and pink mudstones of the Jurassic

Entrada Formation, ~100 m (330 ft) thick. Above the Entrada Formation and beneath the Tropic Shale (Fig. 5A), the Dakota Sandstone (Cretaceous) consists of a sequence, ~50 m (165 ft) thick, of yellow-tan sandstone and conglomerate. Above the Straight Cliffs Formation is a thick (~200 m [660 ft]) sequence of gray to tan sandstone and interbedded shale composing the Wahweap Sandstone (Upper Cretaceous).

To the east of the Kaibab uplift is the Kaiparowits Plateau, which is stratigraphically a broad expanse of mainly Upper Cretaceous strata, namely, the Straight Cliffs Formation, the Wahweap Sandstone, and the Kaiparowits Formation, which is ~400 m (1,320 ft) thick. On Canaan Peak, south of the Table Cliff Plateau (Fig. 2B), the Kaiparowits Formation is overlain by sandstones and conglomerates of the Canaan Peak Formation (~160 m [528 ft]; uppermost Cretaceous and Paleocene) and the Pine Hollow Formation (Paleocene). These formations occupy a Laramide syncline that probably was overlain unconformably by Claron Formation (middle Eocene) (Fig. 5A). At Table Cliffs, Claron Formation is exposed and appears to rest in slight angular discordance upon Kaiparowits Formation. This is the angular unconformity that reveals that Laramide folding in this region ended by the middle Eocene.

East of the Straight Cliffs and west of the Escalante River (Fig. 2B) there is a low desert strip of uppermost Navajo Sandstone, Carmel Formation, and Entrada Formation, creating a predominantly red, rolling landscape. To the north, the Navajo Sandstone is broadly exposed in the region between Escalante and Boulder. To the east of the Escalante River is the Circle Cliffs uplift (Figs. 2, B and D), so named because of deep erosion into the center crest of the uplift, creating a huge elliptical amphitheater (~35 km by ~15 km) bounded by inward-facing cliffs. Stratigraphic units exposed within the Circle Cliffs are Triassic and Jurassic, namely the Chinle Formation, the Wingate Sandstone, and the Kayenta Formation (Fig. 5B). The eastern flank of the Circle Cliffs uplift is the Waterpocket Fold, along which there are fine exposures of Triassic to Cretaceous strata. Although the Wingate through Entrada sequence remains essentially the same as that which is observed to the west, the Upper Jurassic and Cretaceous stratigraphy is distinctively different (Fig. 5B). In particular, the Entrada Formation is overlain by Curtis Formation (~60 m [198 ft]; Jurassic), Summerville Formation (~100 m [330 ft]; Jurassic), Morrison Formation (~160 m [528 ft]; Jurassic), Dakota Sandstone (~120 m [396 ft]; Cretaceous), Mancos Shale (~800 m [2640 ft]; Cretaceous), and the lower part of the Mesaverde Group (~500 m [1,650 ft]; Upper Cretaceous) (see Hintze, 1988). This basic stratigraphy extends eastward into the Henry Mountains and northward to the San Rafael Swell, though in the San Rafael Swell there are exposures as deep as the Kaibab Limestone (Permian) (Fig. 5B).

The mechanical stratigraphy

It is useful from a structural geology perspective to frame the uppermost Paleozoic, Mesozoic, and Eocene strata in terms of a mechanical stratigraphy (Fig. 5C). Three types of mechanical units can be distinguished: stiff, stiff-soft, and soft, where "stiff" can be thought of as "very competent" and "soft" can be thought of as "very incompetent."

The *stiff* units are Kaibab Limestone, the Shinarump Conglomerate Member of the Chinle Formation, Wingate Sandstone, Navajo Sandstone, and Dakota Sandstone. The *stiff-soft* units are Moenkopi Formation, Kayenta Formation, Entrada Formation, Straight Cliffs Formation, Wahweap Sandstone, and Claron Formation. The *soft* units are Chinle Formation, Carmel Formation, Tropic Shale, and Kaiparowits Formation.

Characteristic mechanical responses of these stiff, stiff-soft, and soft mechanical units are best seen in areas of significant deformation, such as along monoclines or where cut by major faults. Consider, for example, the stiff units. Kaibab Limestone is a thick (~250 m [825 ft]), medium-bedded limestone that deforms in broad flexures, e.g., monoclines and anticlines, but is never seen to deform by outcrop-scale folding. The bedded nature of the Kaibab Limestone lends itself to flexural slip. Kaibab Limestone is typically very well jointed and locally is brittlely faulted. Slickenlines on faults in Kaibab Limestone are rarely seen. Tectonic stylolites are completely absent, though at one location along the Kaibab uplift, spaced cleavage was evident. The Shinarump Conglomerate Member is a thin (~20 m [66 ft]), massive, stiff, brittle unit that tends to deform by faulting and relatively widely spaced jointing. Fault surfaces in the Shinarump Conglomerate Member tend to be beautifully polished and slickenlined. Wingate Sandstone is a thick (~130 m [429 ft]), massive sandstone unit quite capable of folding broadly into large anticlines, synclines, or monoclines, but exhibiting only brittle behavior at the scale of outcrops or cliff faces. Systematic closely spaced near-vertical jointing is abundantly developed in the Wingate Sandstone. Deformation bands can develop in the Wingate Sandstone where strain is high, for example along steeply dipping parts of monoclinal folds (Swanberg, 1999). Navajo Sandstone is a very thick (~700 m [2,310 ft]), massive, stiff sandstone unit that deforms into broad folds and giant monoclines. Also, widely spaced, long, continuous joints characterize the Navajo Sandstone. At the outcrop scale and larger, deformation bands and zones of deformation bands are abundantly developed where the Navajo Sandstone has been folded or faulted. Even flexural-slip folding of the Navajo Sandstone is accommodated by the development of layer-parallel deformation bands. Dakota Sandstone is a relatively thin (~50 m [165 ft]), massive, stiff sandstone and conglomerate unit that normally deforms by jointing and faulting, though deformation bands occur locally in especially porous parts.

All of the stiff-soft units are highly vulnerable to flexural-slip deformation when folded because they tend to be very well bedded and marked by alternating competent and incompetent beds (Fig. 5C). Moenkopi Formation is a thick (~520 m [1,716 ft]), thin- to medium-bedded, stiff-soft alternating sequence of siltstones, mudstones, and sandstones, which lends itself to the development of minor folding. Jointing is closely spaced in the Moenkopi Formation. Faulting in the Moenkopi Formation tends to refract in passing from stiff to soft interbeds. Kayenta Formation is a thick (~190 m [627 ft]), medium- to thick-bedded, stiff-soft formation of alternating sandstone and shale. Jointing is relatively widely spaced within the thicker sandstone layers in the Kayenta Formation. Faulting is not uncommon in sandstone layers within the Kayenta Formation where deformed by large folds or faults. Entrada Formation is a relatively thick (~100 m [330 ft]), stiff-soft unit consisting of several thick, massive sandstone members that alternate with several thick, thin- to medium-bedded claystone and mudstone members. Because the alternating incompetent and competent members are very thick, there is disharmony of structural style within the formation. In areas of strong deformation, the soft members may be extremely thinned through faulting, shearing, and even the development of crude cleavage, whereas the thick members may have essentially retained their normal thicknesses and may have responded to the deformation by faulting, deformation banding, and wide-spaced jointing. Thus, scale of observation may dictate whether the Entrada Formation is generally classified as a stiff-soft mechanical unit or is instead subdivided into a number of soft and stiff layers that are treated independently, as if they did not belong to a common stratigraphic formation. The Straight Cliffs Formation and Wahweap Sandstone are both very thick (~320 m [1,056 ft] and ~200 m [660 ft], respectively) and consist of thick-bedded, stiff-soft, alternating interbeds of sandstone and shale. Some of the sandstone beds within these formations have internally deformed through the development of deformation-band shear zones, but normally the response was by faulting and development of slickenlined surfaces. In areas of strong deformation, interbeds of shale and mudstone tend to develop an enhanced bedding-parallel fissility. The Claron Formation (~210 m [693 ft]) consists of two members, the Pink Member and the White Member. Lithologies in the Pink Member are dominantly siltstone, cobble conglomerate, and minor limestone. Bedding is poorly expressed, and thus the Pink Member appears massive. Lithologies in the White Member are limestone and minor siltstone. The Claron Formation does not appear to be especially strong, mechanically, for the siltstones readily responded to deformation by shearing and faulting, and the limestones locally accommodated the development of spaced cleavage and stylolites. Conglomerate lenses and layers may tend to fortify the Claron Formation with some strength, as does the limestone.

The soft mechanical units display thinning in high-strain zones of folds and fault zones (Fig. 5C). The Chinle Formation is a thick (~160 m [528 ft]), thin- to medium-bedded, soft assemblage of clay and mudstone, with local clastic channel deposits and petrified logs. In some places in large monoclinal folds, the Chinle Formation is significantly reduced in thickness, a thinning achieved by flexural flow accommodated by development of shear zones. Carmel Formation is a thick (~210 m [693 ft]), thin- to medium-bedded, soft sequence of interbedded mudstone, limestone, anhydrite, and gypsum. Mudstone is cleaved in regions of high strain, whereas the thin limestone beds typically are fractured or faulted. Alex Bump (personal commun., 1998) has observed that limestone beds in Carmel Formation in the San Rafael region have plentiful stylolites. It is the evaporites that impart a softness to the Carmel Formation so extreme that it is a site for decollement. The Page Sandstone Tongue, where present, can impart some stiffness to the Carmel Formation. For example, along the East Kaibab monocline, the Carmel Formation contains two "tongues" of Page, each ~20 m (66 ft) thick and consisting of massive cross-bedded eolian sandstone. The dominant deformation mechanism within the Page Sandstone Tongue is formation of deformation bands and zones of deformation bands. The Tropic Shale is moderately thick (~140 m [462 ft]) and consists of dark gray to black shale with local interbedded coal seams. The coal deforms by the development of coal cleats and faults (with slickenlines). The shale becomes crudely cleaved and is marked by brittle to semibrittle shear zones. The mechanical softness of the Tropic Shale is revealed by its profound changes in thickness in high-strain zones. Kaiparowits Formation is a very thick (~400 m [1,320 ft]) yet soft unit consisting of shale and mudstone with interbedded siltstone and sandstone. I have not observed its deformational response to full measure, for the Kaiparowits Formation does not occur within any of the Colorado Plateau monoclines. Its likely response would be that of the Tropic Shale.

Deformation bands and zones of deformation bands are most abundant within just four of the Triassic-Eocene strata within the southern Utah parks and monuments region: the Navajo Sandstone, the Page Sandstone Tongue of the Carmel Formation, massive sandstone within the Entrada Formation, and local porous sandstone units within the Straight Cliffs Formation.

Space-satellite views of regional relationships

It is quite revealing to examine expressions of the geology of parts of the Colorado Plateau region of southern Utah in false-color space-satellite imagery (Fig. 6). The false-color images capture the distribution of geologic formations and the locations of the major regional structures. Figure 6A encompasses much of the Kolob Terrace area, showing the systematically jointed Navajo Sandstone in the Zion National Park region. The trace of the Sevier fault is also quite evident in Figure 6A. The Sevier fault controls Long Valley and cuts through Coral Pink Sand Dune State Park. The White Cliffs, formed by Navajo Sandstone, show up especially clearly. Figure 6B encompasses the northern end of the Kaibab uplift, the East Kaibab monocline, and the Kaiparowits Plateau. Navajo Sandstone (white) once again stands out prominently. Figure 6C reveals the expression of the Waterpocket Fold along the eastern margin of the Circle Cliffs uplift as well as the domes produced by laccolithic intrusions in the Henry Mountains and at Navajo Mountain. Figure 6D reveals the form of the San Rafael Swell and also the fault traces that cross it.

BASIN AND RANGE STRUCTURES

Basin and Range faulting has encroached into the southwestern margin of the Colorado Plateau. In particular, three

Figure 6 (this and following three pages). False-color space-satellite views of the region. Yellow hues correspond to red rock; purple hues correspond to gray rock. A: The Kolob region, featuring jointed Navajo Sandstone at Zion National Park and the trace of the Sevier fault near the location of Coral Pink Sand Dunes State Park. B: View of the Kaibab uplift, the trace of the East Kaibab monocline, and the Kaiparowits Plateau. C: View of the Circle Cliffs uplift, the Waterpocket Fold, and the Henry Mountains. D: View of the San Rafael Swell. Images provided by Chevron.

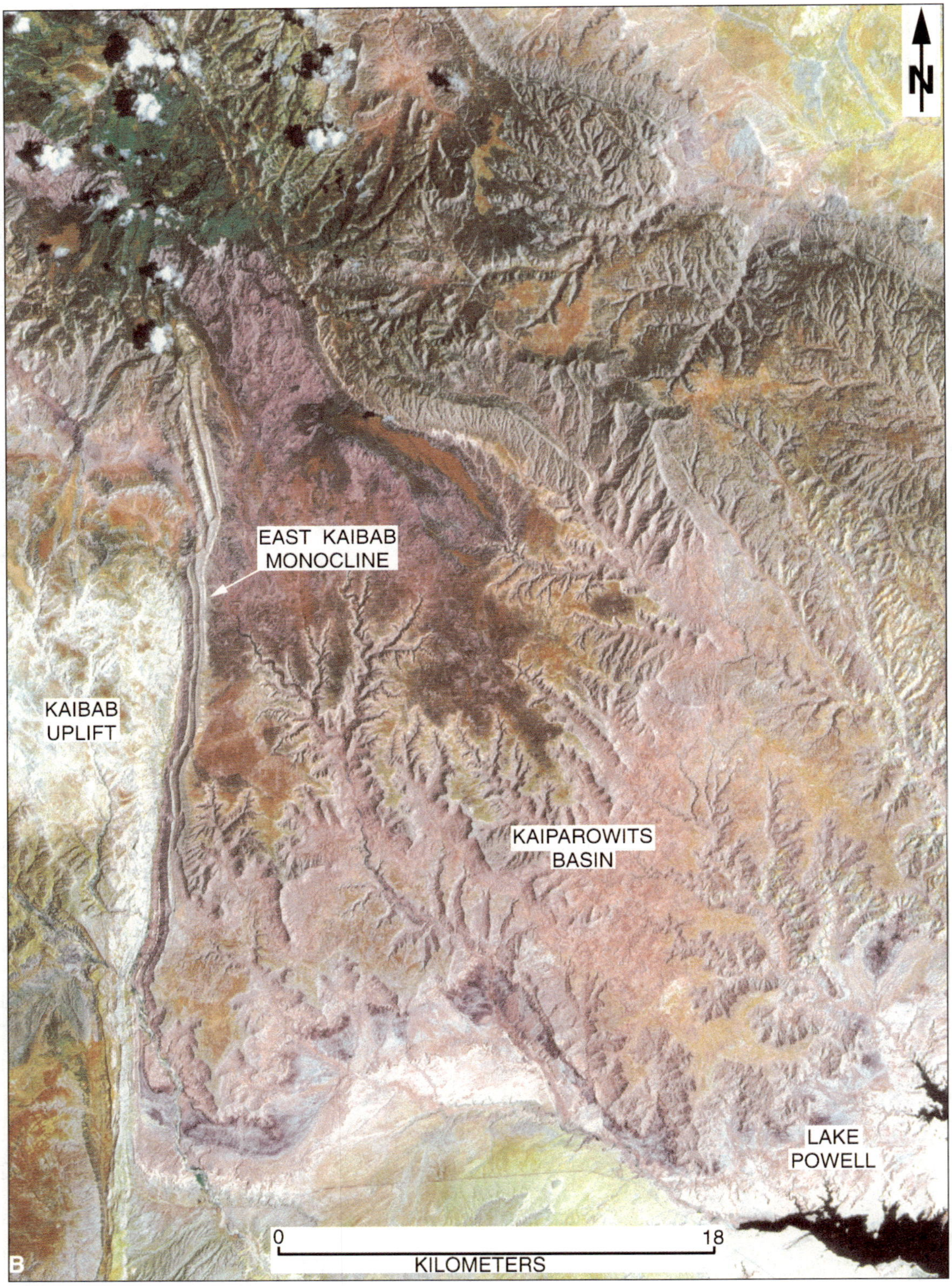
N
EAST KAIBAB
MONOCLINE
KAIBAB
UPLIFT
KAIPAROWITS
BASIN
LAKE
POWELL
0
18
KILOMETERS
B

HENRY
MOUNTAINS
WATERPOCKET FOLD
CIRCLE CLIFFS UPLIFT
C

major Basin and Range faults occur within the region of study: the Hurricane fault, the Sevier fault, and the Paunsaugunt fault (Fig. 2, A and C).

Hurricane fault

The Hurricane fault is an active fault zone that marks the western margin of the Colorado Plateau tectonic province. Projecting northward from the Utah-Arizona state line along a strike of ~N10°–20°E, and dipping steeply westward, the Hurricane fault extends for a distance of ~80 km (50 mi) as a discrete zone before breaking up into a broad area of distributed faulting within Miocene volcanic rocks of the Marysvale volcanic field (Fig. 2, A and C). To the south of the Utah-Arizona state line, the Hurricane fault zone can be traced ~80 km (50 mi) as a discrete structure along a strike direction of ~N15°E. The full length of the Hurricane fault zone may be as great as ~250 km (150 mi) (Stewart and Taylor, 1996).

The Hurricane fault is a normal fault with nearly purely dip-slip motion (Stewart and Taylor, 1996). Beautiful exposures of polished, slickenlined, and grooved fault surfaces testify to the predominantly dip-slip displacement (Fig. 7). Stewart and Taylor (1996) reported that the total stratigraphic separation on the Hurricane fault in southwestern Utah is as much as ~2,520 m (8,316 ft); they documented as much as ~450 m (1,485 ft) of stratigraphic separation for displacement of a Quaternary basalt.

It is not clear when Basin and Range faulting first occurred along the Hurricane fault zone, but on the basis of regional tectonic relationships, the faulting was probably initiated in late Miocene, perhaps at 15 to 12 Ma. Seismic records confirm that

Figure 7. Road cut exposure of the Hurricane fault, near La Verkin, Utah. Note the dip-slip slickenlines and grooves.

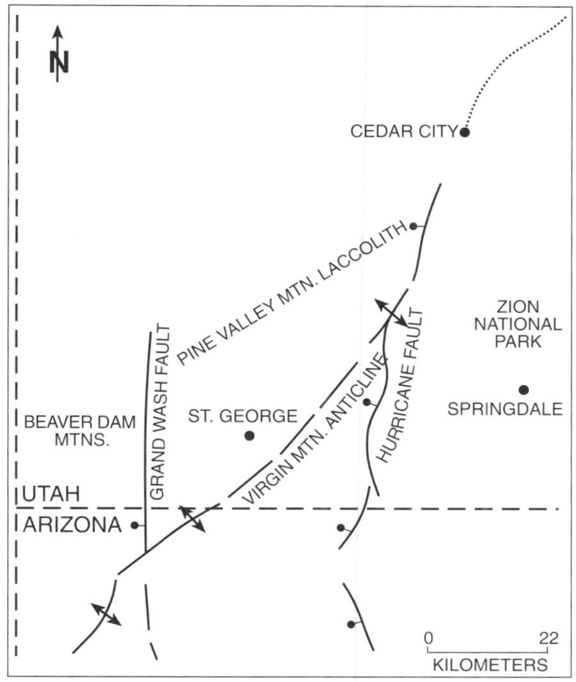

Figure 8. Map showing the geographic relationship of the trace of the Virgin Mountain anticline to the Hurricane fault.

the Hurricane fault remains active today. In fact, the compilation of earthquakes from 1900–1985 in Nevada and Utah by Engdahl and Rinehart (1991) and as illustrated by Smith and Arabasz (1991, Fig. 1, p. 186) reveals very heavy seismic activity along the Hurricane fault. First-motion focal mechanisms for the September 2, 1992 ($M_w = 5.8$), earthquake near St. George record perfect normal, dip-slip motion (Stewart and Taylor, 1996).

The trace of the Hurricane fault coincides with a zone of long-lived tectonic importance and complexity. The Hurricane fault zone is part of the expression of the "Wasatch line," which follows the rifted continental margin of western North America as it existed in the late Precambrian. To the east are the shelf and platform Paleozoic sedimentary rocks of the Colorado Plateau. To the immediate west are miogeoclinal Paleozoic assemblages. The Wasatch line also corresponds roughly to the eastern margin of thin-skinned folding and thrusting. The superposition of tectonic features is striking. For example, the Virgin Mountain anticline, whose form is most clearly expressed just west of the Hurricane fault, can be traced north-northeastward to where both it and the Hurricane fault are coincident with one another, along a trend of ~N10°–20°E (Fig. 8).

Sevier fault

The trace of the Sevier fault zone lies ~65 km (40 mi) east of the Hurricane fault (Fig. 2, A and C). The Sevier fault is a west-dipping high-angle normal fault zone that strikes ~N30°E along a trace length of ~100 km (60 mi), before losing its discrete character within the Miocene volcanic rocks to the north. South of the Utah-Arizona state line, the Sevier fault can be traced to the Grand Canyon, a distance of ~100 km (60 mi).

Normal dip-slip motion has occurred on the Sevier fault, as revealed by slickenlines and by patterns of deformation-band shear zone. Total normal displacement along the Sevier fault ranges from ~600 to ~760 m (2000 to 2500 ft). Geologic mapping of the Sevier fault zone northward from the Utah-Arizona state line reveals that the displacement along it is arranged in interesting ways involving steps, ramps, and relays (Fig. 9). From the Coral Pink Sand Dunes State Park northward to the vicinity of Mount Carmel Junction, the main normal-fault expression is ~N10°E, west side down (Fig. 9). It dips ~75°W. The stratigraphic separation is obvious, because flat-lying Jurassic Navajo Sandstone to the east of the fault is exposed at exactly the same elevation as Upper Cretaceous Straight Cliffs Formation to the west of the fault. The hanging-wall block of the Sevier fault zone is marked by a system of ~N20°W-striking splay faults in the

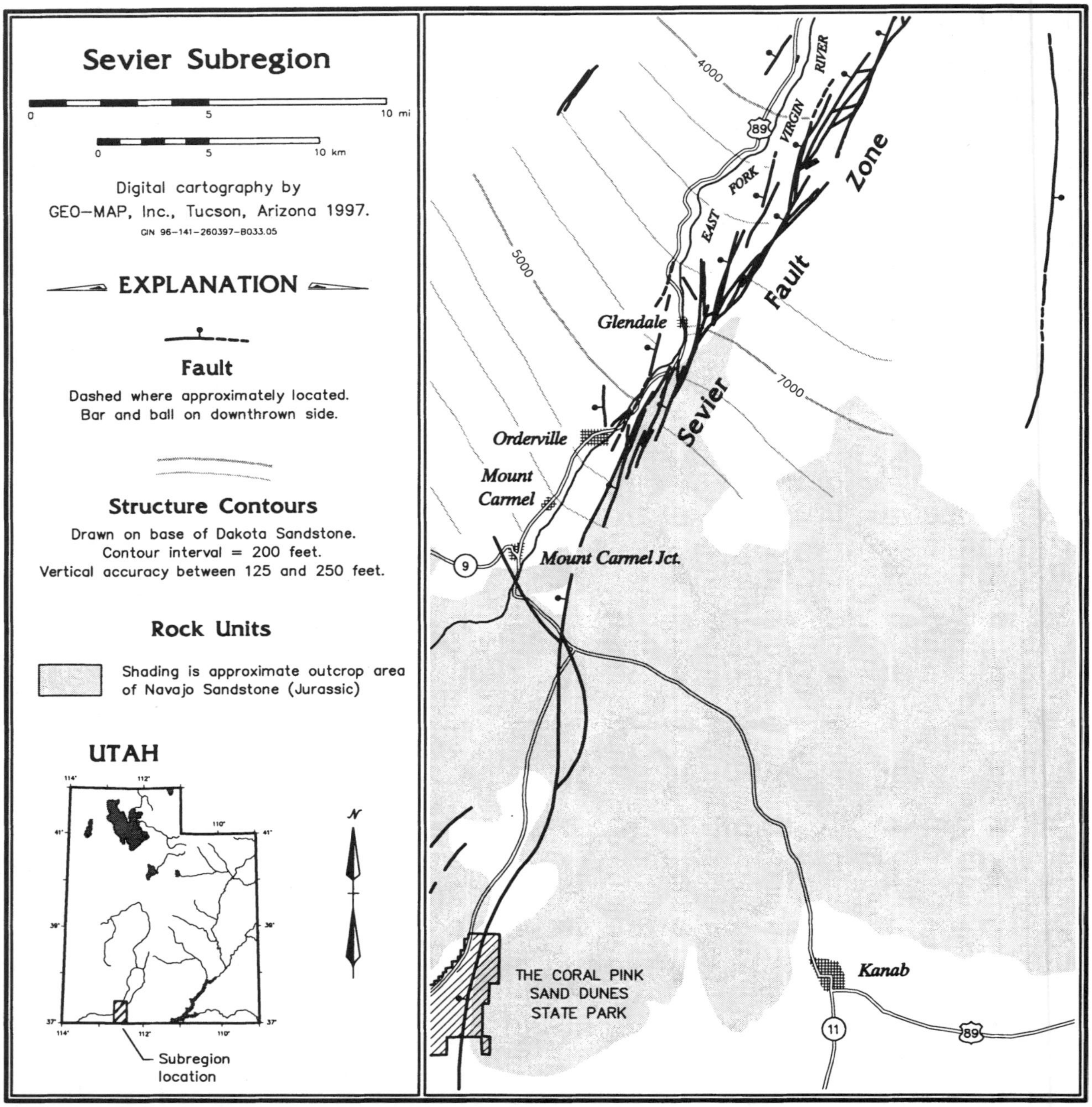

Figure 9. Structure map of the Sevier fault zone, based on a compilation of the mapping of Sargent and Philpott (1987) and Cashion (1961, 1967). Note that contour lines are structure contours.

vicinity of Mount Carmel Junction, some of which are kilometers in trace length.

One of the N20°W splay faults is very well exposed in Carmel Formation (Jurassic) just to the north of Mount Carmel Junction, as described by Higgs and others (1991) and as illustrated in Figure 10. At the northern end of the outcrop there are two main faults, subparallel to one another, which dip ~50°. They have a combined displacement of ~12 m (40 ft). Interbedded sandstone and shale on the hanging wall display rollover, such that over a distance of ~7 m (23 ft), the dip increases from 10° (the regional dip) to 26°, suggesting that the major faults are listric. The crest of the rollover is marked by a graben structure, which is in turn bounded on its southern side by a series of antithetic faults. Along one of the major synthetic normal faults is an excellent example of shale smear (Fig. 11), where sheared hanging-wall mudstones were plastered onto footwall rocks during the faulting.

Just north of Mount Carmel, the Sevier fault zone as a whole

Figure 10. West-directed view of northwest-striking splay faults associated with the Sevier fault zone. The location is 1.6 km (1 mi) north of Mount Carmel Junction on Highway 89. Hanging-wall strata are marked by rollover. Bill Higgs is positioned between the traces of two normal faults, both dipping left (southwest).

Figure 11. Shale smear (dark thin line) along one of the splay faults at the Mount Carmel locality. Geologist is Bill Higgs.

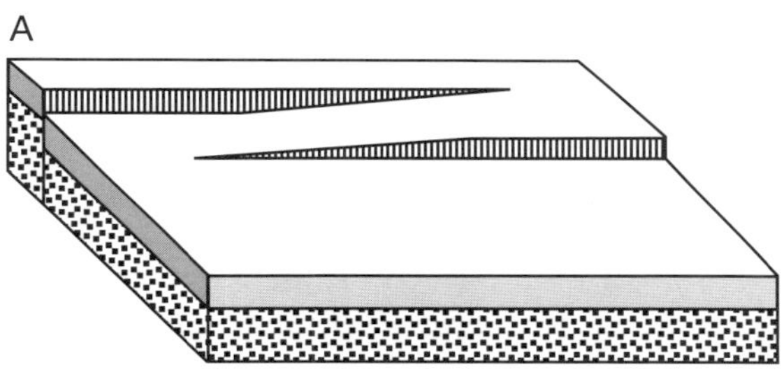

Figure 12. A: Block diagram of relay ramp between right-stepping normal faults. Adapted from Peacock and Sanderson (1994, Fig. 12, p. 159). B: Northeast-directed photograph of expression of relay ramp (dipping sandstone layers) between two overstepping splays of the Sevier fault zone (not shown in the photo). The dip is anomalously steep compared to regional. The dipping strata in the relay ramp are in the footwall of one of the overstepping faults and in the hanging wall of the other. White lines show dip of bedding.

Figure 13. Northeast-directed view, from Highway 12, of the west-dipping Sevier fault zone near the mouth of Red Canyon. Hanging-wall Quaternary basalt (Qb, ca. 500 ka) is downfaulted against footwall Claron Formation (Tc, Eocene). Note the abrupt termination of Claron Formation bedrock against the fault. White lines emphasize bedding and the trace of the fault.

strikes ~N30°E (see Fig. 9), although many of the individual faults strike ~N15°E. This pattern is especially well developed in the Orderville-Glendale region, where the normal-fault segments are right stepping and overlapping, in a way that matches descriptions of relay ramps by Peacock and Sanderson (1994) (Fig. 12A). In these overstepping regions, displacement is transferred from one fault segment to the other by the combination of normal faulting and the formation of relay ramps (Fig. 12A). The relay ramp at Orderville is quite well exposed and is marked by Navajo Sandstone that dips ~15°N (Fig. 12B).

The segmented nature of the Sevier fault continues to the latitude of Long Valley Junction (Fig. 9), north of which the fault has been mapped as a single trace for ~60 km (35 mi). In fact, the fault zone probably extends farther north into Miocene volcanic rocks of the Marysvale volcanic field (Figs. 2C and 9). Near Long Valley Junction, the Sevier fault is exposed at a high stratigraphic level, i.e., within Cretaceous and Eocene strata. An excellent exposure of the Sevier fault can be seen at the mouth of Red Canyon (north side) along Utah Scenic Highway 12, where late Quaternary basalts (ca. 500 ka) are faulted against Claron Formation (Eocene) in the footwall (Fig. 13).

In regional view, strata in the hanging-wall side of the Sevier fault zone are seen to have been broadly dragged by the faulting. This effect shows up clearly in structure-contour patterns of the Mesozoic strata (Figs. 2C and 9). The homoclinal attitude of the Mesozoic strata in the Zion Canyon region is ~N25°W, 2°NE. Yet within several kilometers of the trace of the Sevier fault zone, the strata become drag folded into a syncline that plunges ~1° to the N10°E. The Sevier fault is considered to be late Tertiary to Holocene (Smith and Arabasz, 1991), on the basis of mapping, compilation, and analysis by Anderson and Miller (1979). The probable onset of movement on the Sevier fault is 15 to 12 Ma, consistent with regional timing relationships for Basin and Range faulting. Only two earthquakes appear to have been recorded along the main length of the Sevier fault in Utah between 1900 and 1985, according to the compilation of Engdahl and Rinehart (1991) and as reported by Smith and Arabasz (1991, Fig. 1, p. 286). Nevertheless, these earthquakes suggest that the fault is still active, an interpretation supported by the abundant seismic activity evident near the branch point between the Hurricane fault zone and the northernmost end of the Sevier fault zone, as reported in the Engdahl and Rinehart (1991) compilation.

Paunsaugunt fault

The Paunsaugunt fault is the easternmost of three major Basin and Range faults that occur in the Colorado Plateau region of southern Utah (Fig. 2, A and C). It represents the easternmost expression of Basin and Range faulting in the southern Utah sector of the Colorado Plateau. The trace length of the Paunsaugunt fault is ~66 km (40 mi), along a strike direction of N5°–10°E. The fault dips westward, bordering the eastern margin of the Paunsaugunt Plateau, whereas the Sevier fault, also west dipping, borders the western margin. Part of the significant structural control for the location of Bryce Canyon National Park is the Paunsaugunt fault (Fig. 14), which marks the eastern edge of the fantastic "hoodoo" exposures in the Claron Formation. Creation

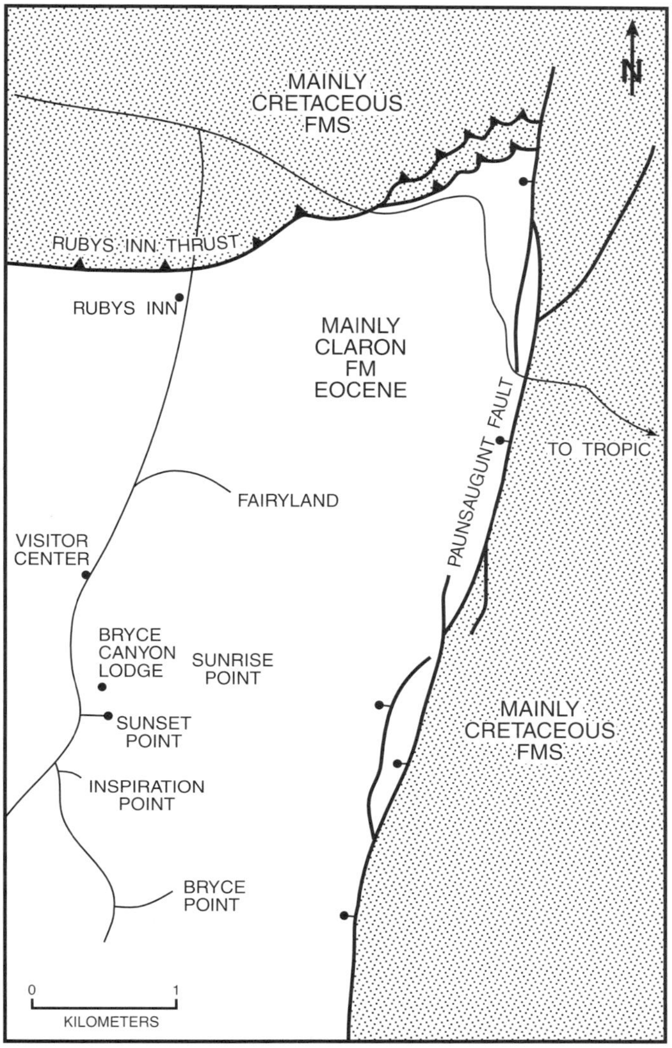

Figure 14. Simplified geologic map of part of the Bryce Canyon region, showing the location of the Paunsaugunt fault (Basin and Range) and the Rubys Inn thrust (Miocene). Adapted from Bowers (1991).

Figure 15. Northeast-directed photographic view of the hoodoos in Claron Formation at Bryce Canyon National Park.

of the hoodoo geomorphology in the Claron Formation in Bryce Canyon (Fig. 15) was favored by headward erosion by tributaries in the Paria drainage system (Fig. 2B), which worked their way north-northwestward through easily eroded Jurassic and Cretaceous formations in the footwall of the Paunsaugunt fault, eventually entering into hanging-wall Claron Formation from below. In effect, the headward drainages of the Paria River system encroached on the Claron Formation from a low topographic vantage on the east, thereby enhancing the formation of slot-like canyons and creating significant undercutting. A part of the hoodoo morphology is controlled by jointing, but the morphology is also controlled by outcrop-pervasive conjugate–thrust-slip and conjugate–strike-slip fault surfaces in the Claron Formation, produced during middle Tertiary (Miocene) thrusting.

In the vicinity of Bryce Canyon National Park, the Paunsaugunt fault strikes ~N15°E and dips ~75°W (Fig. 14). The Claron Formation (Eocene) on the west is separated from Straight Cliffs Formation (Upper Cretaceous) on the east (Fig. 16). Stratigraphic separation is ~600 m (2,000 ft). One can gain a rough sense for the magnitude of this displacement by comparing the elevation of Claron Formation on the west side of the fault at Bryce Canyon National Park (2,100 m [7,000 ft]) with the elevation of the Claron Formation well to the northeast of the fault at Table Cliff (3,120 m [10,300 ft]) (Figs. 2, C and D; 17).

Wherever it is observed, the Paunsaugunt fault zone is marked by gouge and breccia, and the immediately adjacent bedrock is highly fractured and broken. Slickenlines rake ~75°S at localities close to Utah Scenic Highway 12, disclosing a very modest left-lateral component to the predominantly normal movement. To the south, the Paunsaugunt fault divides into two splays, which then rejoin again (see Bowers, 1991). Farther to the south, the Paunsaugunt fault appears to die out in the Carmel Formation (Jurassic), just north of the outcrop trace of the top of the Navajo Sandstone (Fig. 2C). To the north, the Paunsaugunt fault is structurally replaced by the Paunsaugunt monocline, which faces westward. This is a "Basin and Range monocline" that simply appears

Figure 16. A: North-directed photograph of the Paunsaugunt fault, taken from a location ~3 km (1.8 mi) west of Tropic. Hanging wall (left) is Claron Formation (Eocene). Footwall is Straight Cliffs Formation (Upper Cretaceous). B: Closer view of the fault. Note person walking along the skyline ridge to left of largest tree.

to represent a drape step of the Claron Formation along a stretch where the fault displacement has decreased to the point where it can be accommodated by folding.

Frank Royse (personal commun., 1987) has speculated that the Paunsaugunt fault may not die out in the way that the map pattern of Figure 2C suggests. On the basis of seismic reflection records, Royse has suggested the possibility that the Paunsaugunt fault is listric, becoming shallower in its westward dip with depth and merging into the flat-lying, mechanically soft evaporites of the Carmel Formation. Given such a geometry, it is possible that the Paunsaugunt fault is a breakaway for a low-angle detachment-like fault that carries a part of the strata of the Paunsaugunt Plateau westward toward the traces of the Sevier and Hurricane faults. Testing this hypothesis through mapping, I was able to confirm that the Paunsaugunt fault gradually decreases in its westward dip as it cuts deeper stratigraphically from north to south. The lowest dip that I measured was ~35°W at the hanging-wall stratigraphic level of the upper part of the Carmel Formation. The 35°W dip is fully 40° shallower than the dip of the fault at the hanging-wall stratigraphic level of the Claron Formation. However, in following the outcrop exposures of evaporites in the Carmel Formation across the southern margin of the Paunsaugunt Plateau, I was not able to observe any compelling evidence for layer-parallel shearing of the kind that might be consistent with the detachment hypothesis.

As in the case of the Sevier fault, the Paunsaugunt fault is considered to be late Tertiary to Holocene (Anderson and Miller, 1979; Smith and Arabasz, 1991), although the exact timing of onset of faulting is not known. The youngest mapped formations cut by the Paunsaugunt fault are ca. 20 Ma (Bowers, 1991). The Paunsaugunt fault appears to represent the easternmost encroachment of Basin and Range faulting into the Colorado Plateau. The Paunsaugunt fault does not appear to be active today, for no earthquakes were recorded along the fault during the period 1905–1985 (Engdahl and Rinehart, 1991). Furthermore, there is negligible seismicity east of the Paunsaugunt fault (Arabasz and Julander, 1986). The delicate balancing of rocks in the hoodoos of Bryce Canyon might suggest that there have been no recent large earthquakes along the Paunsaugunt fault.

MIDDLE TERTIARY (MIOCENE) STRUCTURES

There are two dominant middle Tertiary (Miocene) structural systems within the region of study. The Paunsaugunt thrust system crosses the northern part of the Paunsaugunt Plateau and involves the Claron Formation (Eocene) in thrust faulting and fault-propagation folding (Fig. 2, C and D). The other Miocene regional structural system comprises the laccoliths and domed Mesozoic strata of the Henry Mountains and Navajo Mountain in the easternmost part of the study region (Fig. 2D).

Paunsaugunt thrust system

The Claron Formation (Eocene) has long been described as flat lying and undeformed, except where affected by Basin and Range faulting (for example, Dutton, 1880; Gregory and Moore, 1931; Gregory, 1951; Bowers, 1972; Hintze, 1988). The Claron Formation (Eocene) postdated Laramide compressional folding and faulting. Thus several observations of contractional deforma-

Figure 17. East-directed photograph of Claron Formation (Eocene) in part of Bryce Canyon National Park. The Table Cliffs, capped by Claron Formation, are in the background. The Claron Formation there resides at a much higher elevation than the Claron Formation at Bryce Canyon in the foreground. Claron Formation is flat lying, except at the "Sinking Ship," where the strata have been drag-tilted into west dips directly adjacent to the the Paunsaugunt fault.

tion in the Claron Formation were quite unexpected (Davis and Krantz, 1986): (1) Discovery of a small (~2- to 3-m offset [~7 to 10 ft]) south-dipping thrust in the Claron Formation directly beneath the Bryce Point Overlook in Bryce Canyon National Park (Fig. 18); (2) recognition of thrust-slip kinematics in cleaved fault gouge in the low-angle fault zone that occurs in Claron Formation along the road cut of Utah Scenic Highway 12 just north of Bryce Canyon National Park (Fig. 14); (3) the presence of well-exposed, steeply dipping to overturned Claron Formation in the northernmost part of Bryce Canyon National Park, just south of Utah Scenic Highway 12 (Fig. 14); and (4) recognition of a thrust relationship captured in a tall (10 m [33 ft]) hoodoo, marked by a caprock of Cretaceous sandstone resting in low-angle fault contact on Claron Formation (Fig. 19).

These observations led Davis and Krantz (1986) to reevaluate the nature of two east-striking faults, mapped by Gregory (1951), that cross the Paunsaugunt Plateau (Fig. 20). Gregory (1951) had mapped these faults as steeply outward-dipping (Basin and Range) normal faults bounding an east-trending horst block (Fig. 20A). The fault to the south is the Rubys Inn fault, sometimes referred to as the Ahlstrom Hollow fault; the one to the

Figure 18. Southeast-directed photograph of Bryce Point showing location of "discovery" thrust in the Claron Formation.

Figure 19. South-directed photograph of tall hoodoo, just south of Highway 12 at the northernmost end of Bryce Canyon National Park. The hoodoo is capped by sandstone of the Straight Cliffs Formation (Upper Cretaceous), which forms a tiny klippe. The remainder of the hoodoo is sheared and faulted Claron Formation (Eocene). The Rubys Inn thrust separates the older-on-younger relationship. Eric Lundin stands at the base of the hoodoo. Note steeply dipping Claron Formation in background.

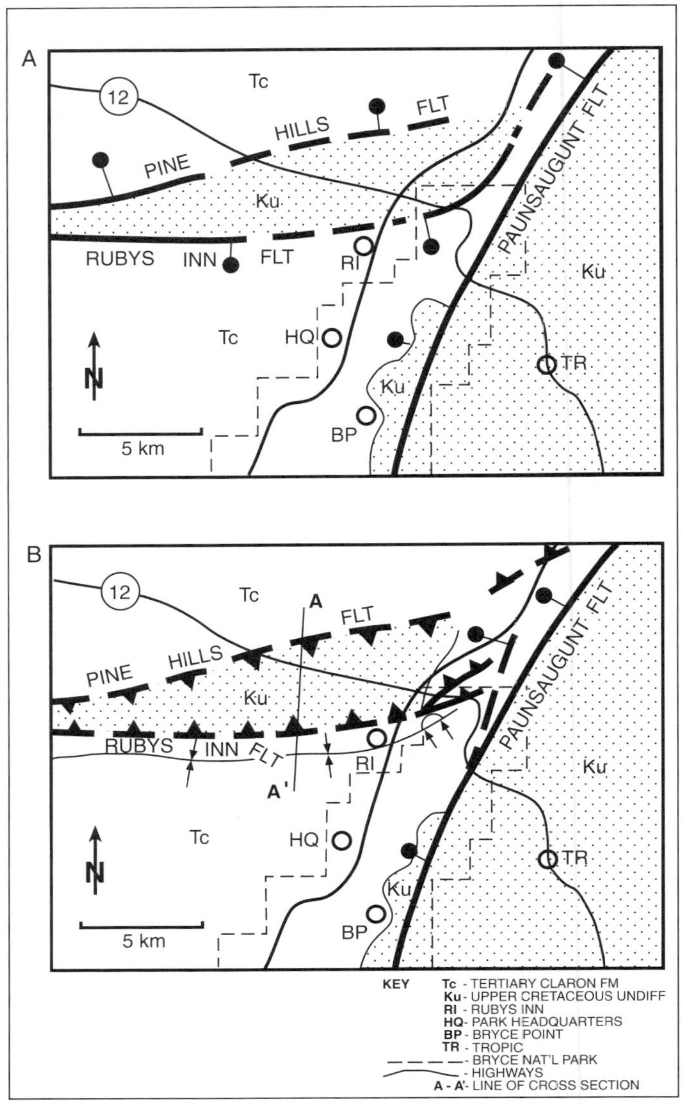

Figure 20. A: Map showing normal-fault interpretations of the east-striking Pine Hills fault and Rubys Inn fault (formerly known as the Ahlstrom Hollow fault), based on the work of Gregory (1951). B: Thrust-fault reinterpretation of these faults based on the work of Davis and Krantz (1986). A–A′ show the location of the cross section in Figure 23.

north is the Pine Hills fault. The Rubys Inn fault was interpreted by Gregory (1951) to be a south-dipping high-angle normal fault, which dropped down a hanging wall of Claron Formation (Eocene) against the Cretaceous footwall. The Pine Hills fault had been interpreted by Gregory (1951) as a north-dipping high-angle normal fault juxtaposing hanging-wall Claron Formation against footwall Straight Cliffs Formation (Fig. 20A). Field observations by Davis and Krantz (1986) led to a new interpretation (Fig. 20B). The Rubys Inn thrust dips ~30°N and places Cretaceous strata over Eocene strata of the Claron Formation. It commonly is associated with large-scale, overturned, south-vergent folding. The Pine Hills thrust dips ~30°S and, like the Rubys Inn thrust, places Cretaceous strata over Eocene strata.

Structural mapping by Lundin (1987, 1989) ensued and resulted in fuller elucidation of the thrust system and in the discovery of magnificent outcrop exposures of the thrust and fault-propagation-fold relationships (Figs. 21 and 22). Lundin (1989) was able to demonstrate that the Rubys Inn thrust extends completely across the Paunsaugunt Plateau from the Paunsaugunt fault to the Sevier fault (Fig. 2B)—a distance of ~20 km (12 mi)—and that the Rubys Inn thrust is itself cut and offset by these two Basin and Range faults. According to Lundin (1989), shortening accommodated by the Rubys Inn thrust ranges from ~350 m (1,155 ft) near the trace of the Sevier fault to ~1,000 m (3,300 ft) near the midpoint of the Paunsaugunt Plateau. Furthermore, Lundin's (1989) mapping revealed that the belt of thrusting turns north-northeastward along the eastern margin of the Paunsaugunt Plateau, extending for a distance of at least 30 km (18 mi) along Johns Valley. There, according to Lundin (1989), the Johns Valley

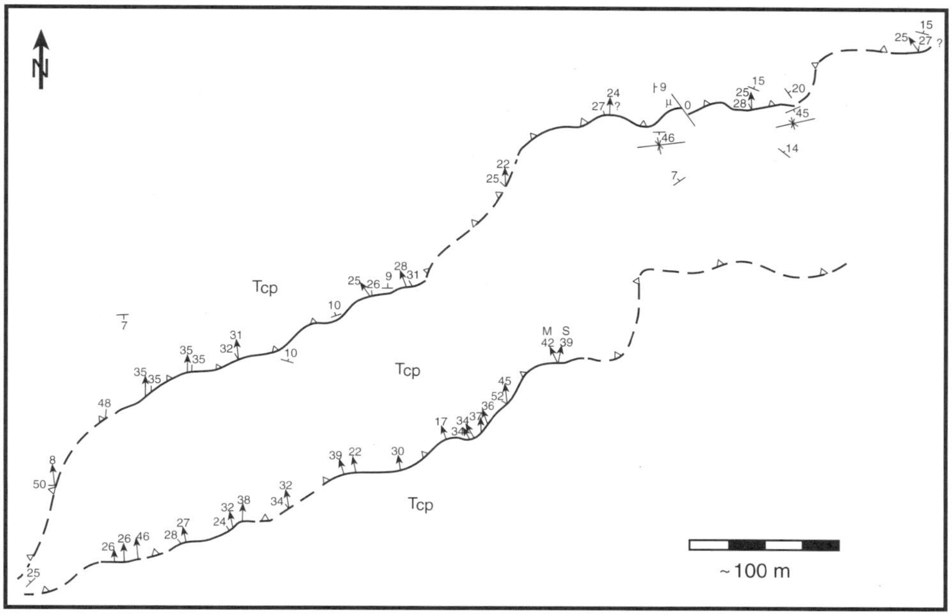

Figure 21. Structural details (field map) of Rubys Inn fault splays just north of Highway 12. Mapping by Eric Lundin (1987). Tcp = Tertiary.

Figure 22. A: Southwest-directed photograph of a splay of the Rubys Inn thrust in Claron Formation. Chuck Kluth is holding up the hanging wall. B: Northwest-directed photograph of a splay of the Rubys Inn thrust in Claron Formation, showing grooves and slickenlines as well as breccia. Two day packs for scale.

thrust has accommodated much of the shortening (~2,000 m [6,600 ft]).

As it turns out, Chevron geologists were aware decades ago of the presence of thrusted Eocene in the Paunsaugunt Plateau! I learned this through Frank Royse (personal commun., 1986), who made available to Lundin and me the results of drilling and seismic reflection studies. These data reveal that the thrust faults dip ~30° into the subsurface, cutting down-section through Claron Formation (Eocene); the Wahweap Sandstone, Straight Cliffs Formation, Tropic Shale, and Dakota Sandstone (Upper Cretaceous); and the Entrada Formation (Jurassic). At the stratigraphic level of the Carmel Formation

(Jurassic), the thrusts flatten into a horizontal, layer-parallel orientation within the evaporites. Thus, Navajo Sandstone is not cut by this thrusting within the Paunsaugunt Plateau region. Seismic reflection profiles further reveal that the Pine Hills thrust is a backthrust.

Lundin and Davis (1987) and Lundin (1989) combined the results of geologic mapping with the subsurface control provided by Chevron, thus creating structure sections that show the nature of the thrusting of Claron Formation in the Paunsaugunt Plateau region (Fig. 23). Sense of thrusting was determined on the basis of fault orientations, fault offsets, overturned folding, and the orientations of grooves and slickenlines (Davis and Krantz, 1986; Lundin, 1987, 1989). Along the northern part of the Paunsaugunt thrust system, where the faults strike north-northeast, the sense of slip is ~S45°E. Along the eastern part of the system, where the faults strike east-northeast, the sense of slip is ~S20°E. Within the main part of the belt, where the thrusts strike east, sense of slip is approximately south to S10°W. Magnitude of offset along the major thrusts appears to range from ~0.5 to 1.0 km (0.3 to 0.6 mi).

The thrusting produced footwall and hanging-wall deformation in the Claron Formation in the form of pervasive small (millimeter- and centimeter-scale displacement) conjugate-thrust faults and conjugate–strike-slip faults, as well as tectonic stylolites, slickolytes, and spaced cleavage in impure carbonates (Davis and Krantz, 1986; Lundin, 1987, 1989). For example, the classic hoodoos and fin-like walls of Claron Formation in Bryce Canyon National Park are marked by conspicuous conjugate thrusts and strike-slip faults (Fig. 24).

The extensiveness of the internal deformational fabric within the hoodoos at Bryce Canyon National Park reveals that the generally accepted interpretation of the formation of hoodoos at Bryce is not complete. The hoodoos are interpreted conventionally as the product of erosion of flat-lying strata cut by vertical joints; the hoodoos emerged as expressions of the vertical jointing and a stratigraphy that offered, from bed to bed, a differential resistance to erosion. It would now appear that the geometries of fault surfaces within the Claron Formation are as important, and locally more important, than jointing in controlling the morphology and orientations of the hoodoos. Furthermore, the differential resistance of bedrock to erosion is not only an expression of the primary lithologies, but also an expression of the differential resistance to erosion of fault products (such as gouge, breccia, and sheared bedrock) achieved through deformation.

The geographic extent of this thrusting was further enlarged by mapping and analysis to the west of the Sevier fault in the Markagunt Plateau (Fig. 25), reported by Nickelsen et al. (1992) and Merle et al., (1993). Although major thrusts are not exposed west of the Sevier fault, the characteristic footwall deformation is quite obvious in the form of outcrop-scale thrust faults, strike-slip faults, and stylolites and slickolytes. The total trace length of the Paunsaugunt thrust system thus appears to be at least 100 km (60 mi). The trace of the thrust system in map view is broadly arcuate (Fig. 25): transport direction is outward to the southeast, south, and south-southwest, always subperpendicular to the trace of faulting. Magnitude of thrust offset, as described above, is comparatively small, on the order of ~300 m (990 ft) to ~2000 m (6,600 ft).

The thrusting postdates deposition of the Claron Formation (Eocene) and probably took place in the interval of ca. 30 to 20 Ma, coeval with the formation of the Marysvale volcanic center located ~40 to 60 km (24 to 36 mi) northwest of the thrust belt. The thrusts and fault-propagation folds associated with the Paunsaugunt thrust system appear to have been produced during gravitational spreading and collapse of the Marysvale volcanic field, which lies to the north-northwest (Fig. 26) (Davis and Rowley, 1993; Merle et al., 1993). The Miocene volcanic rocks rest on Eocene and Oligocene strata, including the Claron Formation (Eocene) and the Brian Head Formation (Oligocene) (Fig. 27). At the time of the Miocene volcanic outpouring the Brian Head Formation and overlying Bear Valley Formation (Oligocene-Miocene) were still relatively unconsolidated and fluid saturated, creating an instability that resulted in mass sliding and the generation of rock-avalanche megabreccias (Anderson, 1993). Among the products is the "Markagunt Megabreccia," which occupies a known mapped area of ~500 km^2 (300 mi^2 (Anderson, 1993). The megabreccia contains blocks

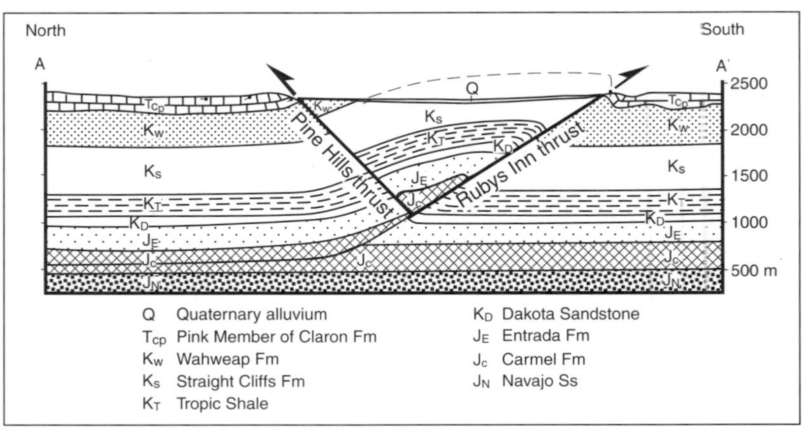

Figure 23. Lundin and Davis (1987) and Lundin (1987, 1989) combined the results of geologic mapping and structural analysis with the subsurface control provided by Chevron, thus creating structure sections that show the nature of the Rubys Inn thrust. Note that the Pine Hills thrust is a back thrust. Location of A–A′ shown in Figure 20B.

that are as large as houses and hotels (Fig. 28). The megabreccia matrix materials are the soft lithologies of the Brian Head and Bear Valley Formations. According to Anderson (1993), these formations were water saturated at the time of loading and collapse. Outcrops thus consist of blocks enveloped in sheared weak lithologies. Sable (personal communication with Anderson, 1993) has mapped slip-line indicators in the Markagunt Megabreccia, and they trend parallel to the orientations of slickenlines in the closest exposures of the Paunsaugunt fault system, namely, south-southwest.

Part of the spreading and collapse of the Marysvale volcanic field may have been favored by the emplacement of laccoliths, like the Spry intrusion, located near the southeast margin of the volcanic field north of Panguitch (Anderson et al., 1990). The laccoliths domed the Claron Formation (Fig. 29). This doming and lateral spreading may have increased the gravitational potential, which in turn contributed to lateral mass movement. At a much deeper level, the gravitational spreading and gliding were accommodated by the low-angle thrusts that sole into Carmel Formation evaporites (Fig. 29), in much the same manner as fault-propagation folds and thrusts actively forming south of the base of Mount Etna today, connected at depth to shallow thrusts that project beneath the base of Etna. The detachment zone there consists of weak bentonitic clays (Borgia et al., 1992).

A two-tiered model of the spreading and collapse of the southern part of the Marysvale volcanic field is thus envisioned (Fig. 29). At the surface, mass wasting and landsliding created the Markagunt Megabreccia, while simultaneously at a deeper level, the entire stratigraphic column above the Carmel Formation evaporites moved as a thin-skinned thrust sheet on the evaporite decollement (Davis, 1997b). Fault-propagation folding accompanied the thrusting.

Figure 24. A: E-directed photograph of thrust fault in a fin-like wall of Claron Formation in the Fairyland Loop area of Bryce Canyon National Park. B: S-directional photograph of a hoodoo (~6 m [zoft] tall) in Bryce Canyon National Park. It is composed of Claron Formation cut by numerous fault surfaces, which are themselves slickenlined. C: SW-directional photograph of well-developed fault surface in Claron Formation along Navajo Loop Trail. The slickenlined fault surface is marked by horizontal slickenlines. The geologists are Drew Davis (#80) and Matt Davis.

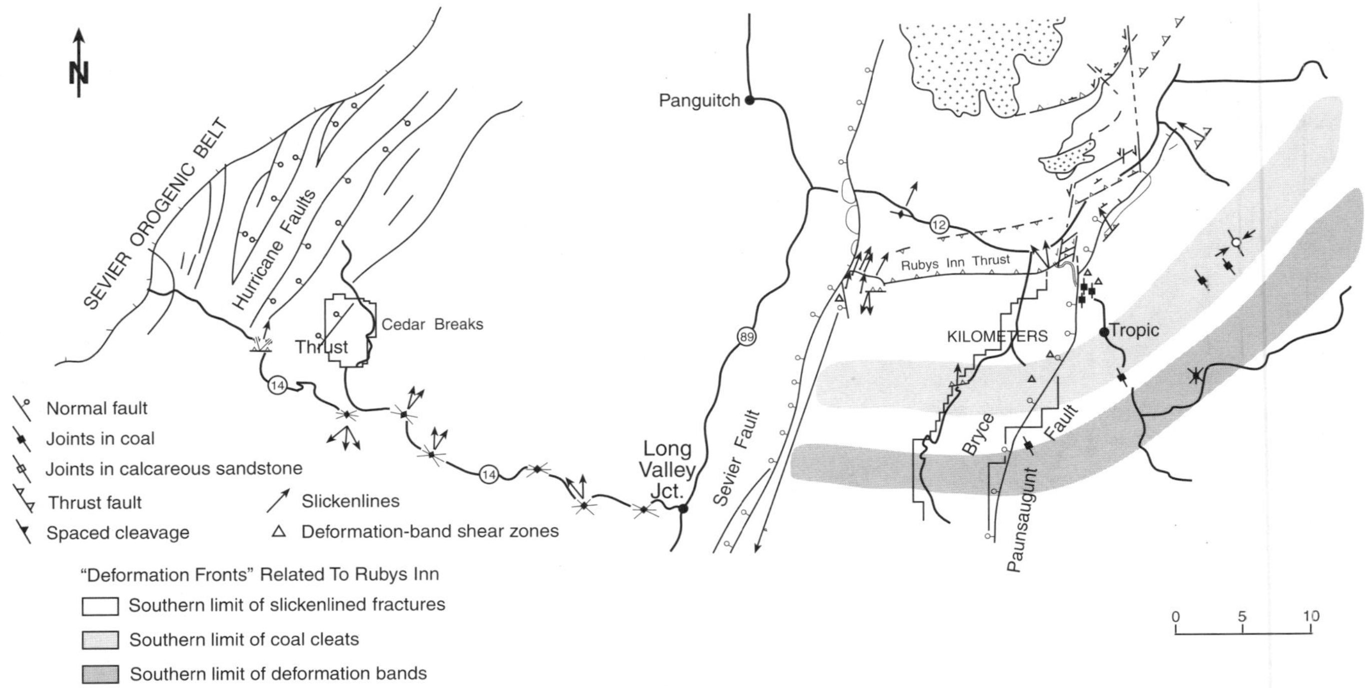

Figure 25. Structure map of the known full extent of the Paunsaugunt thrust belt and related structures. The Rubys Inn thrust and related thrusts were mapped by Lundin (1987, 1989) across the Paunsaugunt Plateau and northwestward along John Valley. Evidence for thrusting west of the Sevier fault was discovered by me, Richard Nickelsen, and Oliver Merle, through study of outcrops of Claron Formation along Route 14 between Long Valley Junction and Cedar Breaks. Nickelsen discovered and mapped the regional presence of deformation fronts relatable to the Rubys Inn thrust (Nickelsen et al., 1992).

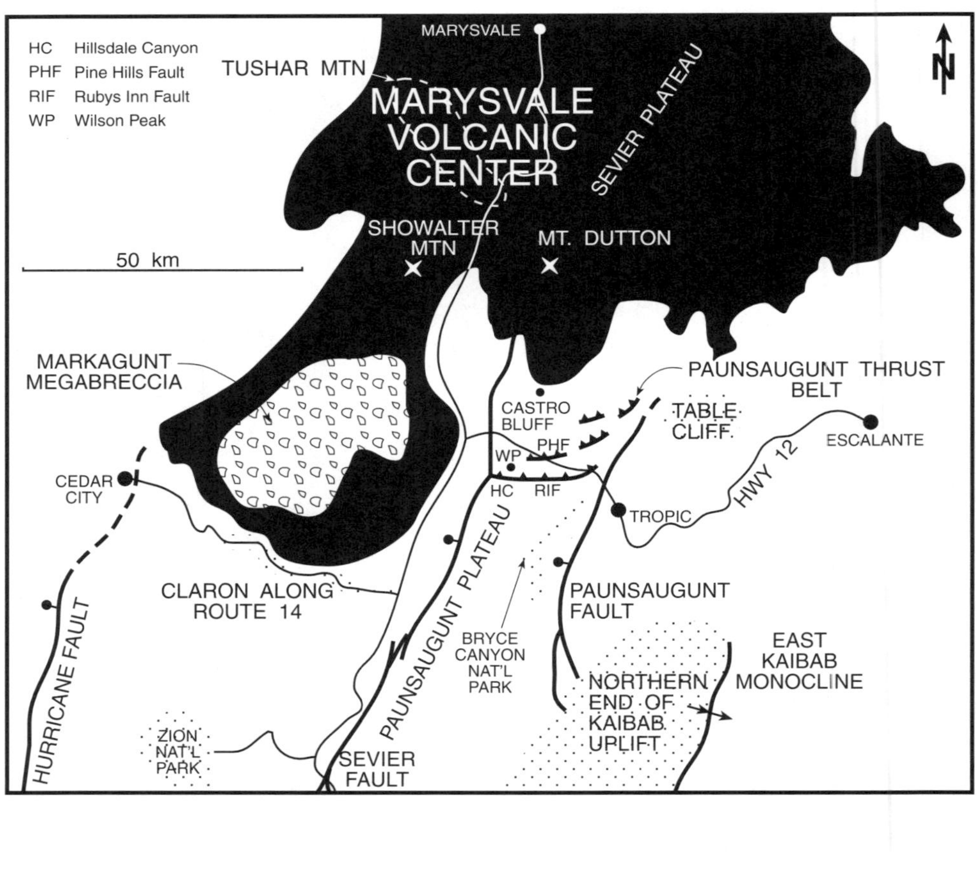

Figure 26. Location of the Marysvale volcanic field in relation to the Paunsaugunt thrust system. From Merle et al. (1993).

Figure 27. The strata that lie just beneath the Marysvale volcanic field are well exposed at Cedar Breaks National Monument. Rocks exposed in the "breaks" (i.e., the cliffs) belong to the Claron Formation. Claron Formation is overlain by Brian Head Formation, which crops out in the dark hill on the horizon in the far distance.

Henry Mountains laccoliths

The Henry Mountains represent the type locality for laccolithic intrusions (Gilbert, 1877) (Fig. 2, B and D). The mountains themselves never appeared on any map until John Wesley Powell's explorations and reports (1875) of the Colorado River region of the West. The Henry Mountains are a field of laccoliths. Made classic by the investigations of G. K. Gilbert (1877), the Henry Mountains were later studied in extraordinary detail by Hunt (1946) and Hunt et al. (1953). Hunt dedicated his career to this igneous tectonic system and to the scientific mind and achievements of G. K. Gilbert (Hunt, 1988a). As one example of Gilbert's genius, Hunt reported that Gilbert correlated the laccolithic development of the Henry Mountains with the igneous activity that was so extensive during the middle Tertiary in the Marysvale volcanic field within the High Plateaus of Utah, 200 km (125 mi) to the west. This interpretation somehow anticipated the results of modern geochronological determinations. On the basis largely of fission-track determinations for zircon and sphene (Naeser, 1978), Sullivan (1998) reported 30 to 20 Ma emplacement ages for the magmatism of the Henry Mountains. Nelson (1998) has refined this interpretation even further through

Figure 28. Exposures of large exotic blocks of Mount Dutton Formation (Miocene) encased within the Markagunt Megabreccia. The two massive outcrops on the skyline are exotic blocks. Smaller blocks are in the foreground.

$^{40}Ar/^{39}Ar$ ages determined for samples collected in the Henry Mountains. He determined that magmatism spanned 31.2–23.3 Ma, i.e., the middle to late Oligocene.

The Henry Mountains contain five laccolithic centers: Mount Ellsworth, Mount Holmes, Mount Hillers, Mount Pennell, and Mount Ellen (Fig. 30). They are aligned along a trend of ~N25°W. The cumulative length of the Henry Mountains is ~67 km (40 mi); and their width is ~23 km (14 mi). The laccolithic intrusions are diorite porphyry and quartz diorite porphyry, which spread as

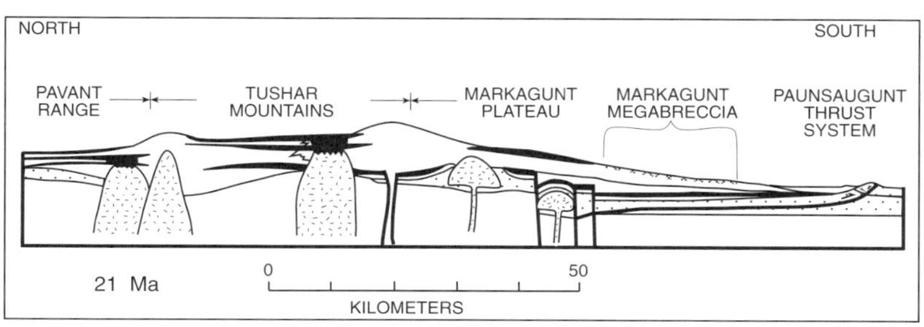

Figure 29. Schematic cross section showing the relationship of the Paunsaugunt thrust belt to the Marysvale volcanic field and the Markagunt Megabreccia.

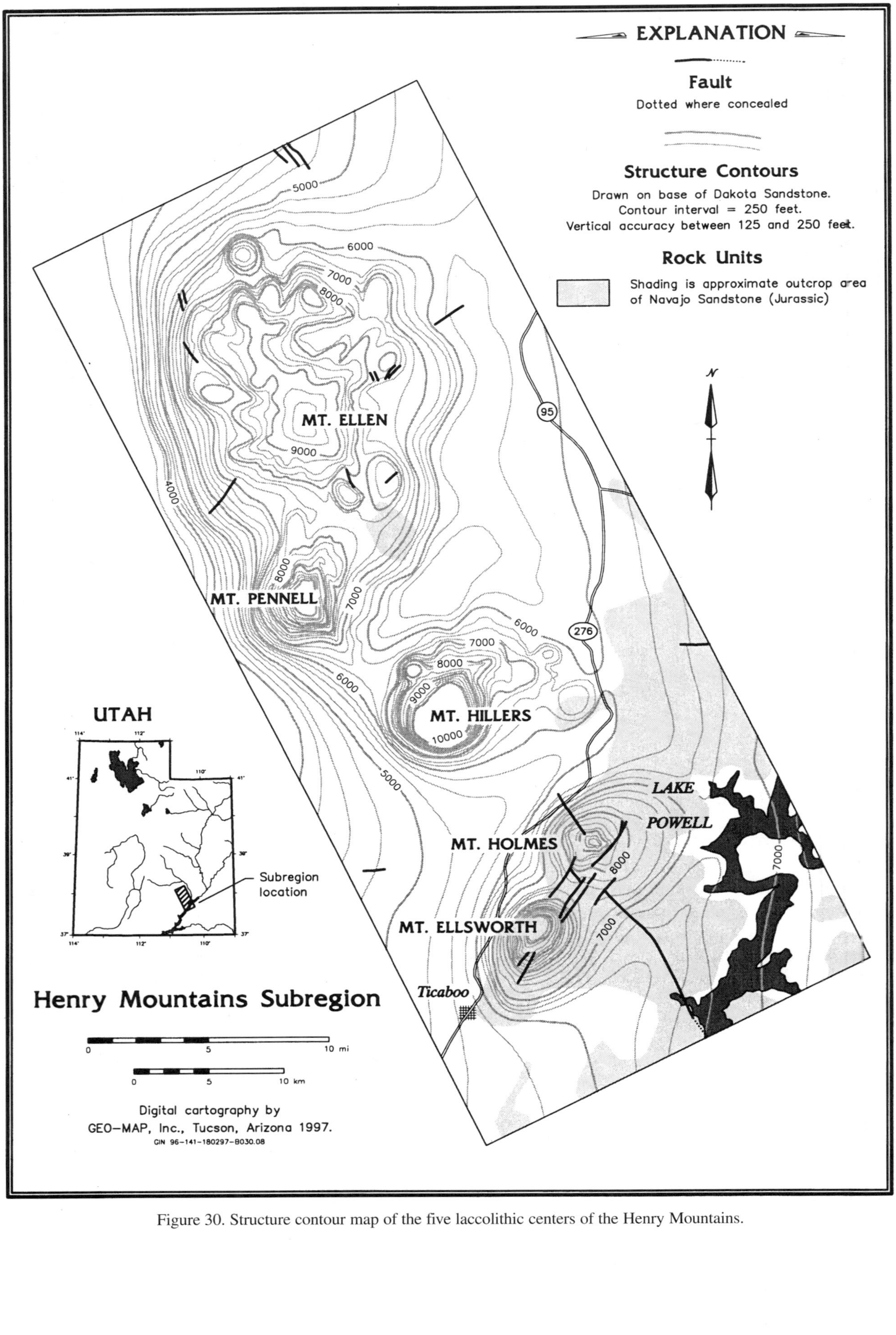

Figure 30. Structure contour map of the five laccolithic centers of the Henry Mountains.

Figure 31. Photographs of two of the laccolithic centers of the Henry Mountains showing darker colored igneous centers and lighter colored Jurassic cover rocks. A: Mount Holmes. Jurassic beds are dipping gently. B: Mount Hillers. Jurassic beds are dipping steeply.

enormous sill-like bodies, at a stratigraphic position in the uppermost part of the Paleozoic section. Overlying Mesozoic formations, forming a sequence ~4 km (2.4 mi) thick, were domed as intrusion proceeded. Jackson and Pollard (1990) and Jackson (1998) pointed out that Mount Holmes, Mount Ellsworth, and Mount Hillers each are ~12 km (7.2 mi) in diameter, but with amplitudes of 1.2 km (0.7 mi), 1.85 km (1 mi), and 3.0 km (1.8 mi), respectively. Where regional bedding approaches each of the laccolithic centers of the Henry Mountains, the bedding becomes rather abruptly flexed and domed. Each center is to some extent breached by erosion, and thus the uparched bedding can be seen circumscribing the base of the laccolithic center (Fig. 31). For example, beds on the flank of Mount Holmes dip (outward) ~20°, beds on the flank of Mount Ellsworth dip ~50° to 55°, and beds on the flank of Mount Hillers dip at ~80° (Jackson, 1998).

Gilbert (1877) viewed the central intrusions as having grown from sills by lifting and bending of the overlying strata. In contrast, Hunt et al. (1953) viewed central intrusions as cylindrical stocks from which emanated small flanking sills. According to Hunt's (1983, 1988b) view, the laccoliths grew laterally as tongue-shaped masses from the flanks of the cylindrical stocks. Geologic mapping and analysis carried out by Jackson and Pollard (1988a, 1988b) led them to support Gilbert's model. For example, the results of the paleomagnetic investigations by Jackson and Pollard (1988a) reveal that the sills and thin laccoliths overlying the central intrusion of each center were indeed emplaced horizontally as proposed by Gilbert.

Jackson and Pollard's (1988a) showed that strata were uplifted ~1,200 m (3,960 ft) at Mount Holmes, 1,800 m (5,940 ft) at Mount Ellsworth, and at least 2,500 m (8,250 ft) at Mount Hillers. Radii of all three of the domes are similar, ~5 to 7 km (3 to 4 mi). The strata over the domes have a double-hinge shape, similar to that of a monocline, with the lower hinge concave up and the upper hinge concave down (Jackson, 1998). Very gently dipping distal parts of the flanks of the domed strata are 3 to 4 km (1.8 to 2.4 mi) long, presumably underlain by sills and minor laccoliths, and they show no evidence for radial shortening.

Jackson and Pollard (1988a, 1990) reported that the flexural doming resulted in the mechanical thinning of the stratigraphic sequence by flexural slip and flexural shear as beds slipped and sheared past one another. A dramatic expression of the result of this flexural shearing occurs on the east flank of Mount Hillers where the bedding surfaces of nearly vertical Mesozoic strata are marked by vertically raking megamullions and megagrooves related to the bedding-parallel shearing (Fig. 32). Jackson and Pollard (1990) emphasized as well the role of circumferential and radial stretching of beds that accompanies the amplification of the domes during continued laccolithic emplacement.

Figure 32. Photograph of "wall" of nearly vertical Navajo Sandstone foreground on the flank of Mount Hillers (left background). In the plane of the bedding are giant megamullions and megagrooves, formed during bedding-parallel shearing as the beds were flexed upward during laccolithic intrusion.

LARAMIDE STRUCTURES

The chief expressions of Laramide tectonics in the region are fold structures: the East Kaibab, Waterpocket, and San Rafael monoclines; the regional, broad, doubly plunging anticlinal uplifts of which they are a part; and numerous lesser anticlines, synclines, and monoclines (Fig. 2D) (Kelley and Clinton, 1960). The monoclines are among the largest and longest uninterrupted folds in the world and probably are either great self-similar fault-propagation folds or progressive limb-rotation folds that link to major regional faults and shear zones at depth. With the chief exception of the Utah part of the East Kaibab monocline, the faults are blind. Neither the San Rafael monocline nor the Waterpocket Fold, at the present levels of exposure, are associated with monocline-contiguous faults. However, the Utah part of the East Kaibab monocline is marked by a length-parallel fault of transpressive character that cuts through the exposed Mesozoic stratigraphy (Davis and Tindall, 1996; Tindall and Davis, 1999). The Waterpocket Fold along the eastern margin of the Circle Cliffs uplift is fault free at surface levels of exposure. Yet there is a significant west-northwest–striking reverse left-lateral fault (the Miners Mountain fault) that is associated with the Miners Mountain uplift immediately north of the Circle Cliffs uplift (Smith et al., 1963; Anderson and Barnhard, 1986; Billingsley et al., 1987; Bump et al., 1997) (Fig. 2, C and D). Laramide displacement along the Circle Cliffs uplift has largely been transferred to the Miners Mountain uplift across the Miners Mountain fault-zone boundary (Bump et al., 1997). On the San Rafael Swell there is a system of normal faults in the arched and crestal region of the structure, which perhaps in part includes faults related to the formation of the uplift in Laramide time (Fig. 2, C and D). However, a significant part of the system of west- to northwest-trending fault zones, which cut obliquely across the full length and breadth of the San Rafael Swell, may be post-Laramide, related to collapse upon dissolution of Jurassic evaporites in the subsurface (Witkind, 1991, 1994).

Kaibab uplift

The Kaibab uplift occupies both northern Arizona and southern Utah (Babenroth and Strahler, 1945). The full length is ~190 km (117 mi), 80 km (50 mi) of which reside in southern Utah (Figs. 2, C and D; 33). The space-satellite photographic expression of the Kaibab uplift is impressive (Fig. 34). The Utah part of the uplift, at its widest, is ~30 km (19 mi) in breadth, broadening southward into Arizona to a maximum width of ~40 km (25 km). Structurally, the Kaibab uplift is a large asym-

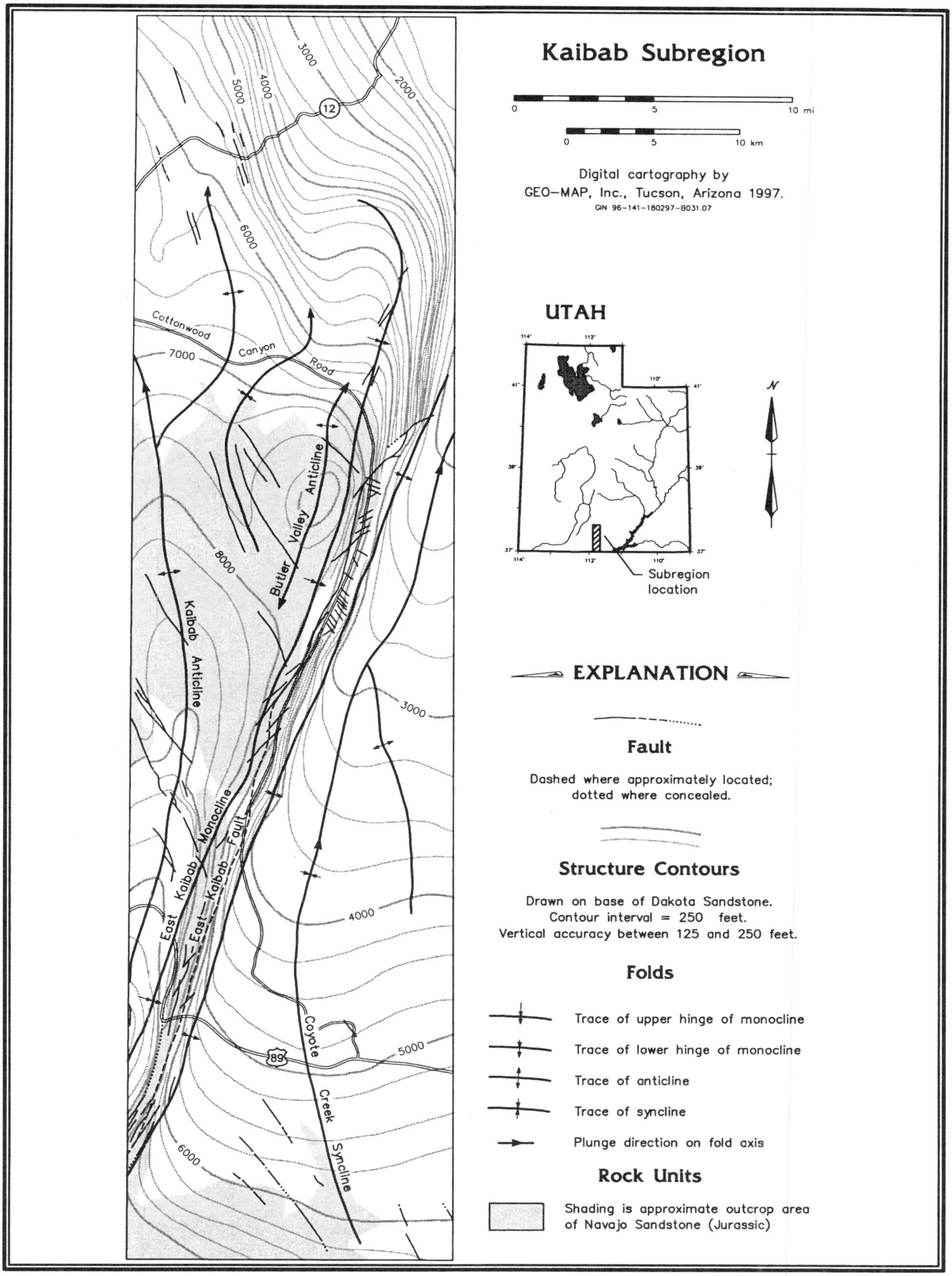

Figure 33. Structure map of the Utah part of the Kaibab uplift and East Kaibab monocline. Note that the contours are structure contours.

Figure 34. False-color space-satellite view of the north-plunging "nose" of the Kaibab uplift. Broad white expanse is the expression of the gently dipping Navajo Sandstone. Steep upturned beds along eastern margin of the uplift are the expression of the East Kaibab monocline. Cretaceous rocks of the Kaiparowits Plateau (purple) lie east of the East Kaibab monocline. (Photograph supplied by Chevron.)

Figure 35. A: North-directed photograph of the East Kaibab monocline. Gently dipping Navajo Sandstone (Jurassic) caps the top of the monocline, on the far left. More readily eroded Carmel Formation, Dakota Sandstone, and Tropic Shale crop out in the foreground. Straight Cliffs Formation is on the far upper right. B: South-directed view of the East Kaibab monocline. Straight Cliffs Formation is in the far upper left. The strike valley is mainly occupied by Tropic Shale. The prominent flatirons are Dakota Sandstone.

metrical anticline, with a long, gently dipping west limb and a moderately to steeply dipping east limb, which is the East Kaibab monocline (Fig. 35). The crest line of the Kaibab uplift is the trace of the Kaibab anticline, located just 5 km (3 mi) west of the upper hinge of the East Kaibab monocline (Fig. 33). The structural relief between the crest of the Kaibab anticline and the upper hinge of the East Kaibab monocline is ~1,500 m (5,000 ft). The Utah part of the East Kaibab monocline trends ~N20°E, parallel to the trace of the Kaibab anticline. Northward it dies out in sympathy with the plunge end of the Kaibab uplift itself (Figs. 33 and 34), at a plunge angle of ~5°. At the very north end, the gently dipping Upper Cretaceous strata are overlain in angular unconformity by flat-lying Eocene strata of the Claron Formation in the vicinity of the Table Cliff Plateau, thus disclosing the Laramide timing of monocline formation. Paleocene formations at the very top of Canaan Peak, southeast of the Table Cliff Plateau, are folded in harmony with Laramide deformation (Bowers, 1972).

Abundant, closely spaced faults cut and displace Jurassic and Cretaceous formations along the northern part of the monocline (Sargent and Hansen, 1982; Davis and Tindall, 1996; Davis, 1997a, 1997b; Christofferson and Davis, 1997; Tindall and Davis, 1999) (Fig. 36). These faults are antithetic left-lateral faults that strike northwestward within a major reverse, right-handed strike-slip brittle shear zone, which can be traced as part of a zone from the northernmost end of the East Kaibab monocline southward to the Utah-Arizona state line. Antithetic faulting is especially well displayed along the Cottonwood Road (Fig. 37A), which for most of its path follows the East Kaibab monocline. Near outcrops known by local ranchers as "Candyland," some of the reverse, left-handed antithetic faults are exposed in the road surface itself, especially when the road is freshly bladed (Fig. 37B). The significant strike-slip component of faulting is expressed in slickenlines and grooves on large, well-defined fault surfaces, such as the one shown in Figure 38. From north to south, this major west-dipping fault cuts down-section from the level of the Upper Cretaceous Straight Cliffs Formation to the Triassic Chinle Formation. The right-handed component of displacement may be on the order of ~5 km (3 mi).

Literature on Colorado Plateau monoclines has commonly emphasized the role of reactivation of Precambrian fault zones (e.g., Huntoon and Sears, 1975; Davis, 1978; Davis et al., 1981; Huntoon, 1993). Huntoon and Sears (1975), on the basis of mapping and analysis, concluded that the East Kaibab monocline roots into a major Proterozoic basement shear zone. This relationship is exposed in the Grand Canyon. It is clear that the fault zone accommodated normal displacement in Proterozoic time. Development of the East Kaibab monocline appears to be related to Laramide northeast-southwest compressional stresses that reactivated, in an oblique-slip manner, the shear zone (Davis and Tindall, 1996; Tindall and Davis, 1999). The concept of reactivating near-vertical Proterozoic fault zones through horizontal compression has, in practice, been fraught with some concerns about how to overcome the high magnitude of resolved normal stress that would arise.

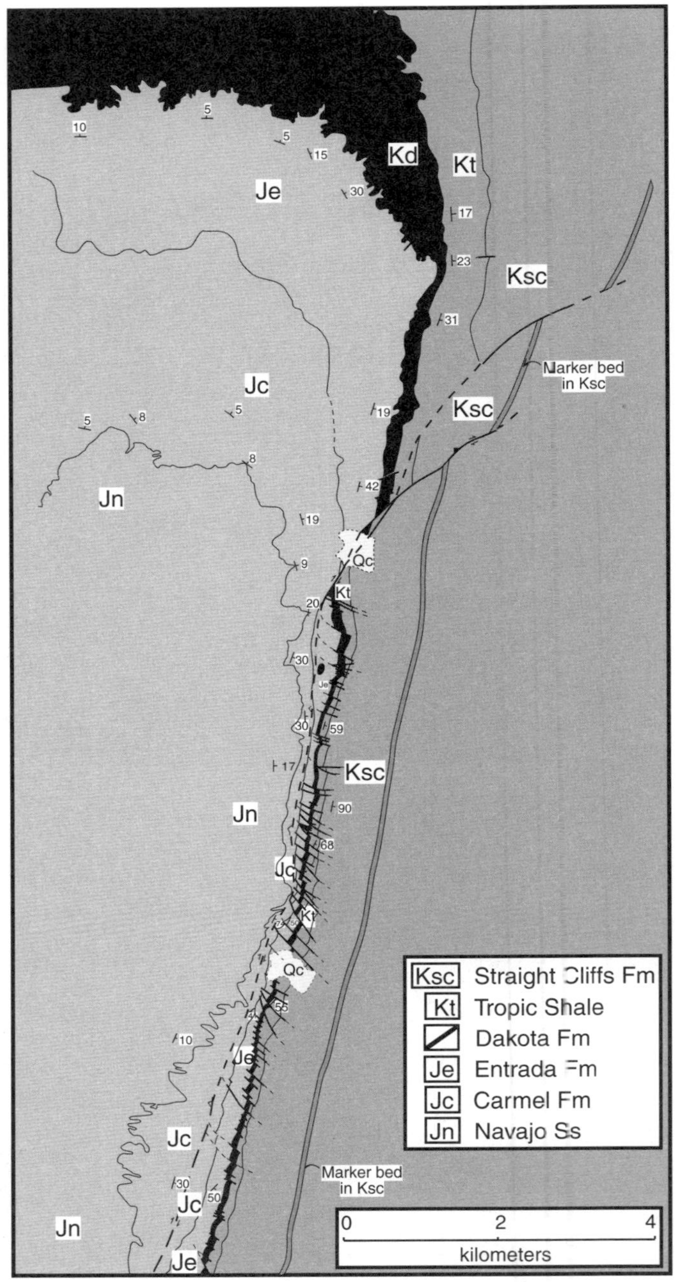

Figure 36. Structural geologic map of a segment of the northern part of the East Kaibab monocline, based on mapping by Sarah Tindall in 1996.

Horizontal compressive stresses acting on vertical faults make it difficult to achieve reactivation (Fig. 39A). This concern largely disappears when the perspective shifts from cross-sectional to map view (Fig. 39B). A N65°E-striking horizontal compressive Laramide stress acting on a N10°E-striking, near-vertical Proterozoic fault zone is ideally suited to reactivate the shear zone in a right-handed strike-slip fashion (Tindall and Davis, 1999).

Figure 37. A: South-directed view of Candyland along Cottonwood Road. White rock on right is the Page Sandstone Tongue of the Carmel Formation. B: Reverse left-handed antithetic faults are in places well exposed in the Cottonwood road surface itself. Here the faults cut and displace units within the Carmel Formation. String-mapping of the relationships is being carried out by Pilar García. This view is southwestward.

Figure 38. West-directed photograph of a reverse, right-handed fault surface, which here separates the Page Sandstone Tongue of the Carmel Formation from the younger redbed units within the Carmel Formation. Note that the surface of the fault is marked by south-raking slickenlines and grooves (white line slanting left). Note also that the younger redbed units strike into the fault zone, rotated by antithetic faults (white line shows trace of bed).

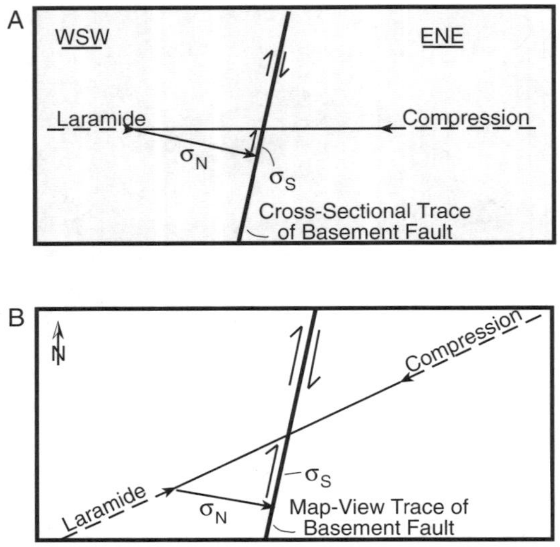

Figure 39. A: Cross-sectional diagram showing the resolved shear stress (σ_S is shear component; σ_N is normal component) on a steeply dipping basement fault, given Laramide horizontal compressive stress. B: Map-view diagram showing resolved shear stress on a steeply dipping basement fault, given N65°E-trending horizontal compressive stress. This configuration favors right-handed transpressive strike slip.

Kaiparowits Plateau

Directly east of the East Kaibab monocline is the Kaiparowits Plateau (Fig. 2, C and D), which in map view is expressed as an elongate triangle of Upper Cretaceous rocks that tapers from a width of ~100 km (60 mi) on the south to a width of ~16 km (10 mi) on the north (Sargent and Hansen, 1982) (Fig. 40). The strata of the Kaiparowits Plateau are deformed into a basin whose axis plunges N20°W at an angle of ~1° or 2°. Long continuous folds in the southern two-thirds of the Kaiparowits Plateau are anticlines and synclines with limb dips of just 2° to 3°. The only notable exception to this rule is the northernmost expression of the Echo Cliffs monocline, a N10°W-trending, east-facing monocline whose trace length in Arizona is ~116 km (70 mi). In southern Utah, this monocline dies out ~17 km (10 mi) north of the Utah-Arizona state line, and it displays a maximum structural relief of ~390 m (1300 ft). Maximum dip of bedding along this monocline in the region of study is ~15°.

The anticlines and synclines in the southwestern part of the Kaiparowits Plateau are arranged in an unusual "dendritic" pattern, resembling a drainage system flowing north-northwestward (Fig. 40). The three main synclinal folds (Coyote Creek, Wahweap, and Last Chance Creek) converge northward into the Table Cliff syncline (Figs. 2, B and D; 40), as if the belt of folds was pinched by indention of the Kaibab uplift from the southwest. The Table Cliff syncline trends N10°W, which is parallel to all of the folds in the northern one-third of the Kaiparowits Plateau. The axial trace of the Table Cliff syncline passes at high elevations through Canaan Peak (2,820 m [9,300 ft]) and Table Cliff Plateau (3,120 m [10,300 ft]); the syncline involves uppermost Cretaceous and Paleocene strata (Sargent and Hansen, 1982).

Canaan Peak itself is very interesting geologically. Bowers (1972) interpreted it as a gravity-slide block composed of Claron Formation. Indeed, the Claron Formation that occupies Canaan Peak is especially highly fractured, jointed, and even fissured, as if it were subjected to faulting at or near the surface. Its structural provenance is not known, but perhaps could be determined by structural analysis of faults, fractures, and fissures at the Canaan Peak locality. In all probability, the Claron Formation now exposed on Canaan Peak may have once resided several kilometers (at most) to the east, having slid down the dip direction of the west-facing Dutton monocline (Figs. 2, B and D; 40).

Some of the folds in the northern part of the Kaiparowits Plateau are very impressive in structural relief. Upper Valley anticline is an asymmetrical fold whose west limb is the west-facing Dutton monocline (Fig. 40). Structural relief on the Upper Valley anticline is more than 1,200 m (4,000 ft). Limb dips on the Dutton monocline are as great as 40°. East of the Upper Valley anticline is the Alvey Wash syncline, a reasonably symmetrical fold with limb dips of as much as 8°. The northeasternmost margin of the Kaiparowits Plateau trends N30°W, parallel to and just west of the Escalante monocline, the western limb of the asymmetrical Escalante anticline (Figs. 2, B and D; 40). Jurassic Navajo Sandstone composes the main surface exposure of the Escalante monocline, striking N25°–30°W and dipping as much as 25°SW (Fig. 41). The trace length of the Escalante monocline does not exceed 33 km (20 mi). It dies out gently to the south-southeast. To the northwest, it is covered by Miocene volcanic rocks and even younger basin-fill sediments. The Miocene volcanic rocks are, in turn, cut and displaced by abundant northeast- to north-northeast–striking Basin and Range normal faults (Fig. 2, B and D), several of which cut across the Escalante anticline in the vicinity of Hell's Backbone (Fig. 42).

South of Escalante, the eastern limit of the Kaiparowits Plateau trends N50°W (Figs. 2, B and D; 40). The eastern margin of the Kaiparowits Plateau is Fiftymile Mountain, an incredibly straight cliff escarpment, 330 m (1,000 ft) high, supported by Upper Cretaceous Straight Cliffs Formation (Fig. 43). Bedding in the Straight Cliffs Formation along this escarpment dips 2°WSW. The structural boundary of the Kaiparowits basin lies east of Fiftymile Mountain, trending parallel to the Escalante River (Figs. 2B and 40). The structural expression of the eastern margin of the Kaiparowits basin is a system of broad, gentle anticlines and synclines, four in all, that plunge very gently southward (Fig. 40). Each of these four folds is parallel to the others and curved in form, trending due north to the north and N50°W to the south. Limb dips are less than 3°.

Circle Cliffs uplift

The next major Laramide structure east of the Kaiparowits basin is the Circle Cliffs uplift, the east margin of which is the Waterpocket Fold (Figs. 2, B and D; 44). Wonderful exposures

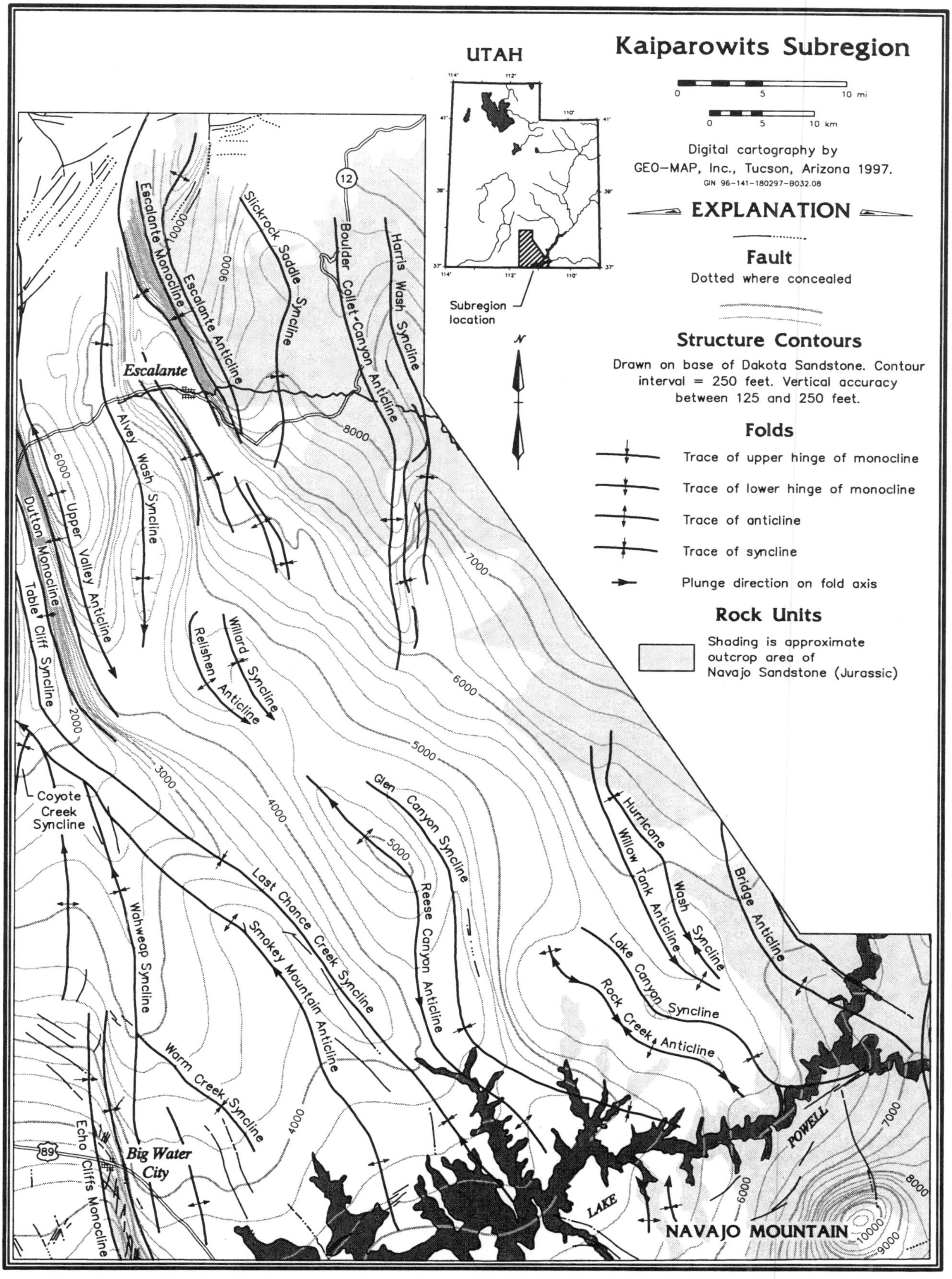

Figure 40. Structure map of the Kaiparowits Plateau, compiled largely from Sargent and Hansen (1982).

Figure 41. North-directed photograph of the Escalante monocline. Navajo Sandstone is dipping west-southwest at ~25°.

occur in Capitol Reef National Park (Smith et al., 1963; Billingsley et al., 1987). The west margin of the Circle Cliffs uplift, which trends ~N40°W, essentially coincides with the Escalante River (Fig. 2, B and D). The Circle Cliffs uplift is entirely contained in Utah and is 140 km (85 mi) in length. The gently dipping west limb ranges in strike from N35°W to N55°W, and dips 3°–4 °SW. The monoclinal eastern limb—the Waterpocket Fold—strikes N20°W and ranges in dip from 15°ENE to 80°ENE. The Waterpocket Fold is spectacular both at ground level (Fig. 45) and from space-satellite level (Fig. 6C). At the widest extent, the Circle Cliffs uplift, including the broad west limb and narrow monoclinal east limb, is ~33 km (20 mi). The uplift achieves this width at the Circle Cliffs proper, which structurally is distinguished by closure of a doubly plunging N30°W-trending anticline, the Circle Cliffs anticline (Fig. 44). Structural relief from the base of the western limb of the anticline to the crest is 900 m (3,000 ft).

Less than 8 km (5 mi) east of the crestal culmination of the Circle Cliffs anticline is the trace of the upper hinge of the Waterpocket Fold, an enormous monocline with maximum structural relief of ~1,666 m (5,500 ft) (Figs. 44 and 45). Directly east of the lower hinge of the Waterpocket Fold is the Henry Mountains syncline (Fig. 44). Strata east of the trace of this syncline dip southwestward toward the Waterpocket Fold at an angle of just 2° to 3°. The N20°W trace of the Waterpocket Fold is slightly discordant to that of the Circle Cliffs anticline (Figs. 2, B and D; 44), such that the traces of these two structures diverge northwestward. To the south-southeast, the Circle Cliffs anticline and the Waterpocket Fold become parallel, and progressively the trend of each bends southward from S10°E to S5°W. The structures project directly toward Navajo Mountain, a dome expressed in Navajo Sandstone and attributable to the presence of a shallow Miocene laccolith (Fig. 2, B and D). The Navajo Mountain dome is ~8 km (5 mi) in diameter and has 600 m (2,000 ft) of structural relief.

In detail, the Waterpocket Fold is marked by minor fluctuations in strike and limb dip, expressed in structure contours as a pinch-and-swell appearance (Fig. 44), but overall the Waterpocket Fold maintains its N20°W trend. In contrast to the East Kaibab monocline that formed obliquely to the direction of Laramide tectonic stresses, the Waterpocket Fold formed essentially at right angles to the direction of Laramide compression.

Miners Mountain uplift

Directly north of the Circle Cliffs Uplift is Miners Mountain, the physiographic expression of Miners Mountain uplift, yet another Laramide uplift (Figs. 44, 46, and 47). Miners Mountain is a faulted, doubly plunging Laramide anticline that trends N55°W. Approximately 40 km (25 mi) in length and 25 km (15 mi) in breadth, this anticline is bounded by high-angle faults that strike N55°–70°W. The faulted fold is marked by structural relief of nearly 1,500 m (5,000 ft) and represents part of the structural relay between the Circle Cliffs uplift on the south and the San Rafael Swell to the northeast (Bump et al., 1997) (Fig. 46). Previously mapped by Smith et al. (1963) and Billingsley et al. (1987), the Miners Mountain uplift is bounded on the southwest by the N65°W-striking Miners Mountain fault and associated monocline. The monocline trends parallel to the fault and verges southwestward. The northeastern backslope of the uplift dips as steeply as 20°ENE, and it is within this backslope that the Visitors Center of Capitol Reef National Park resides.

Miners Mountain is the Teasdale anticline of Anderson and Barnhard (1986). They determined that its southwest margin (their "Cocks Comb monocline") accommodated mainly left-handed strike-slip movement during the Laramide. Their interpretation was based on stress inversion of fault and slickenline data. Bump et al. (1997) concluded that this structure is a reverse, left-handed transpressive structure, with perhaps 1–2 km (0.6–1.2 mi) of left-lateral offset. Prior to this work it was generally understood that the Waterpocket Fold extended well north of the Circle Cliffs uplift, at least as far north as Highway 24 and the Visitors Center (Fig. 2, B and D). Instead, the Waterpocket Fold is *replaced* to the north-northwest by the Miners Mountain monocline in an oblique, left-stepping arrangement (Figs. 44 and 46). One implication is that the east-northeast–dipping Navajo Sandstone exposures that are the "trademark" of the "Waterpocket Fold" in Capitol Gorge and the vicinity of the Capitol Reef National Park Visitors Center are *not* an expression of the Waterpocket monocline, even though described as such on maps and in pamphlets.

The northern margin of Miners Mountain uplift is the Teasdale fault zone, which like the Miners Mountain fault strikes west-northwest (Figs. 44 and 46). This zone is marked both by contractional structures and by extensional structures. The Teasdale fault may represent yet another reverse, left-handed strike-slip fault zone (Anderson and Barnhard, 1986). To the south of the Miners Mountain uplift, in the subregion of "replacement" of the Circle Cliffs uplift by the Miners Mountain uplift, the

Figure 42. Some of the dramatic and rugged landscape of the Hell's Backbone region at the northern end of the Escalante monocline. Bedrock is Navajo Sandstone. Geologists are Jessica Graybill and Sarah Tindall.

Figure 43. A: North-directed aerial view of rocks of the Straight Cliffs Formation (Upper Cretaceous) in the vicinity of Straight Cliffs and Fiftymile Mountain. Photograph by Kurt Constenius. B: Northwest-directed view of the Straight Cliffs from ground level.

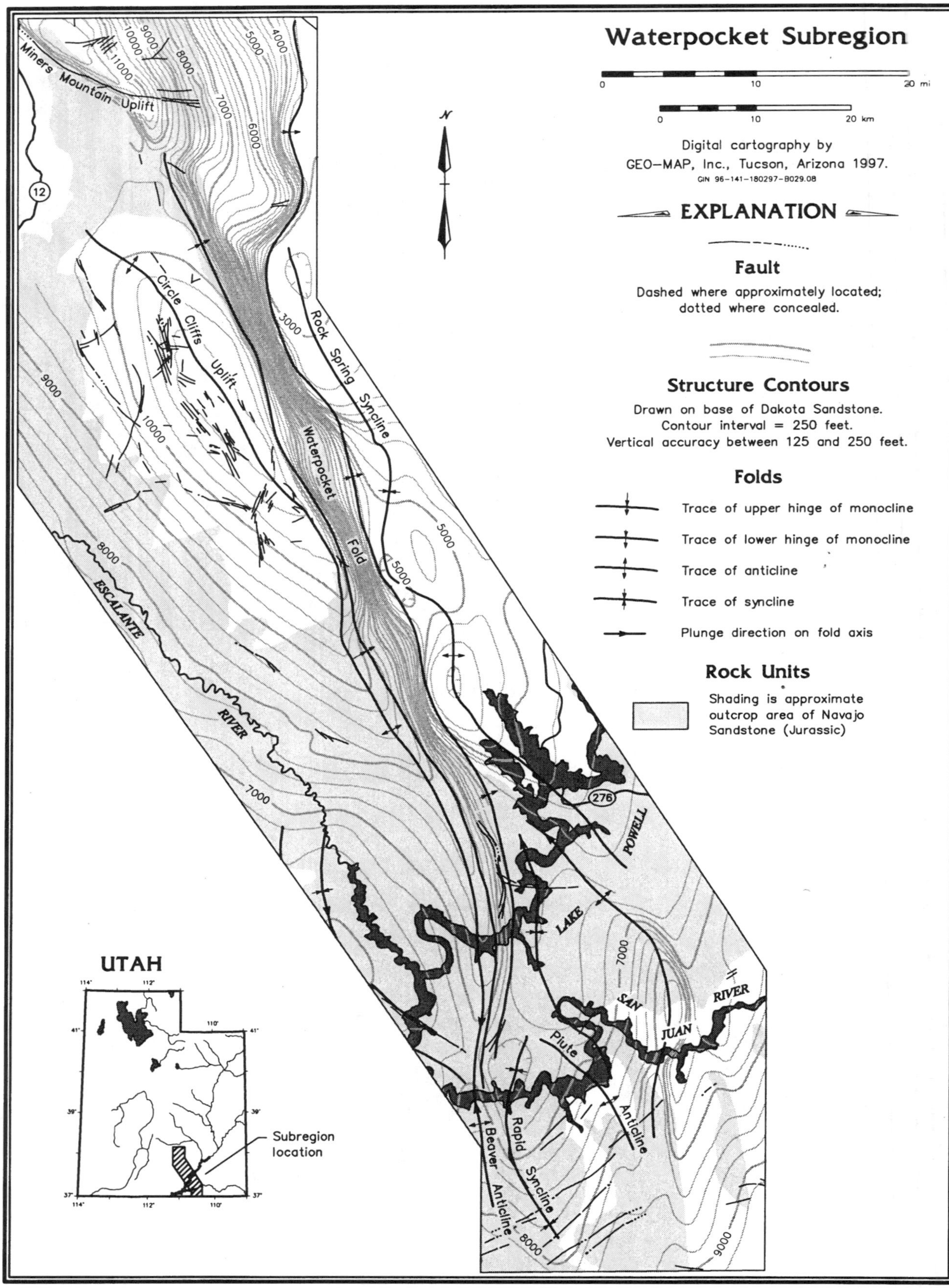

Figure 44. Structure map of the Circle Cliffs uplift region. Note that contours are structure contours.

Figure 45. Southwest-directed view of the Waterpocket Fold. Navajo Sandstone outcrops are massive and white.

Navajo Sandstone abruptly bends eastward and then northward again in a right-lateral jog (Figs. 44 and 47). The width of the jog is ~8 km (5 mi), and strata along this stretch strike east-northeast and dip ~15°S. This jog appears to be a part of the expression of the plunging southeasternmost end of the Miners Mountain uplift.

Deformation along the Circle Cliffs and Miners Mountain uplifts can be explained by east-northeast–west-southwest Laramide compression. Shortening was perpendicular to the trend of the Circle Cliffs uplift and the Waterpocket Fold. The Waterpocket Fold, throughout much of its length, is a fault-propagation fold (or drape fold) associated with a blind thrust or reverse fault at depth, perhaps related to a reactivated basement shear zone. In the vicinity of Miners Mountain, the shortening induced by east-northeast–west-southeast compression imposed transpressive left-slip movements on the west-northwest–striking Miners Mountain fault (and monocline) and perhaps along the Teasdale fault as well.

San Rafael Swell

North-northeast of the Miners Mountain uplift is the San Rafael Swell, yet another of the classic Laramide uplifts (Figs. 2, B and D; 6D; 48) (Gilluly, 1928). The San Rafael Swell is ~150 km (90 mi) in length and 65 km (40 mi) at its greatest width. The entire swell is convex southeastward, curving from N45°E in the south to N10°E in the north (Fig. 48). Furthermore, it is doubly plunging, nosing out at the northern and southern ends. Like the other major Laramide uplifts in the region, the San Rafael Swell is an east-facing asymmetric anticline, marked by a broad, western limb dipping uniformly 3°WNW, and a narrow eastern limb, the San Rafael monocline (Fig. 49), which dips from 15°E to 80°E. The crestal line of the uplift is the San Rafael anticline (Fig. 48), and it lies on average just 5 km (3 mi) west of the monocline. Structural relief on the swell is nearly 2,120 m (7,000 ft). Structural relief on the monocline itself approaches 1,210 m (4,000 ft).

The San Rafael Swell is marked by pervasive faulting systematically oriented obliquely to the trend of the uplift (Fig. 48), as originally mapped by Gilluly (1928). Trace lengths of many of the faults are many kilometers, although displacements are only on the order of meters or 10s of meters. At the southern end of the San Rafael Swell, the faults occur in two sets, the dominant one striking west and a less well developed set striking northwest. Farther to the north, the dominant fault set strikes northwest, and there is a much less well developed set striking northeast. Along the very crest of the San Rafael Swell, north of its culmination, there is an interesting display of faulting in the Chimney Rocks area. Two sets are equally well developed, one striking west-northwest, the other striking east-northeast. Finally, at the very northern end of the San Rafael Swell, faults strike essentially west.

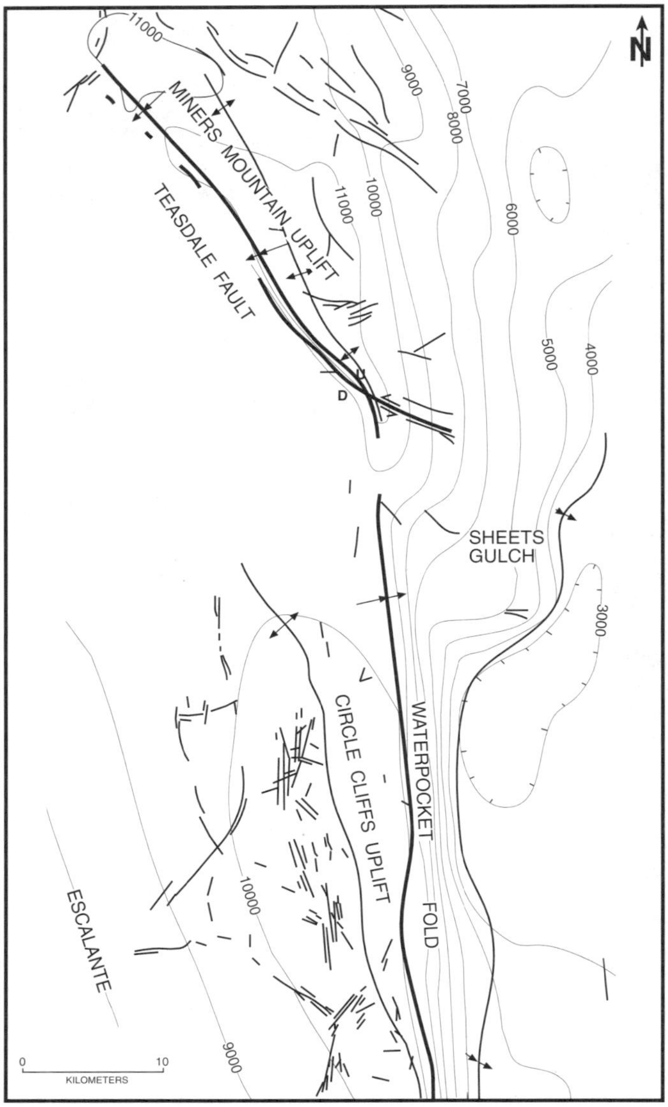

Figure 46. Map showing the relationship between the Circle Cliffs uplift and the Miners Mountain uplift as well as the associated monoclines, i.e., the Waterpocket Fold and Miners Mountain monocline, respectively (Bump et al., 1997).

Figure 47. North-directed aerial photographic view of the Waterpocket Fold along the eastern edge of the Circle Cliffs uplift. In the far distance, the Navajo Sandstone makes a jog to the right (east) at the location where displacement is transferred from the Circle Cliffs uplift to the Miners Mountain uplift. Photograph by Kurt Constenius.

Some of the most conspicuous faulting in the San Rafael Swell, notably that in the Chimney Rocks area, appears to be Laramide. Krantz (1986, 1988), applying the theory of Reches (1978, 1983), demonstrated that the intersecting east-northeast– and west-northwest–striking faults are normal faults that developed in a three-dimensional strain system, in a manner consistent with the doubly plunging nature of the Laramide uplift. Directions of stretching are both east-west and north-south, transverse and parallel to the crest line of the San Rafael Swell at this location. Much of the faulting within the San Rafael Swell, however, has been interpreted to be the result of post-Laramide collapse accompanying the dissolution of Jurassic salt in the subsurface (Witkind, 1991, 1994).

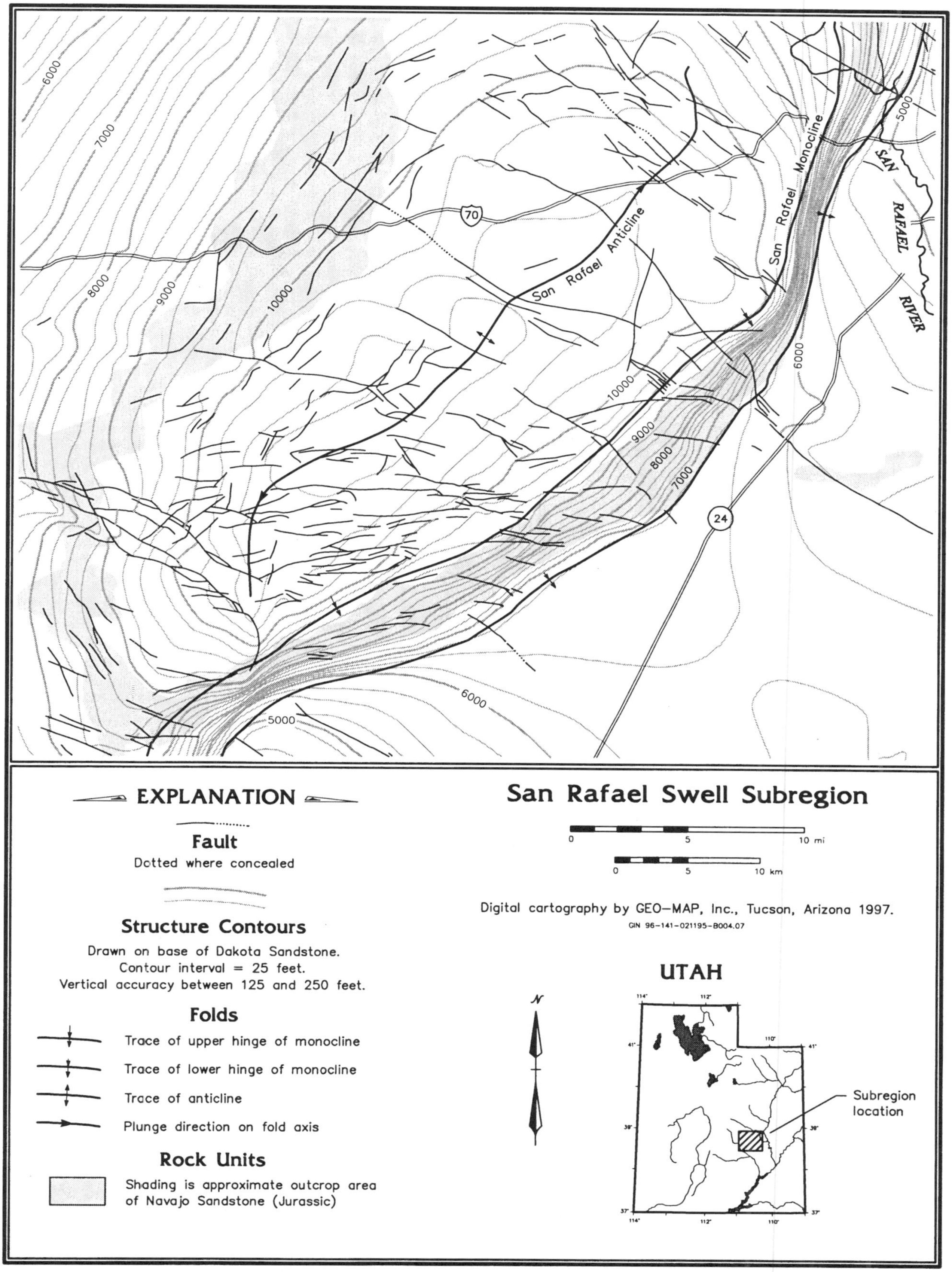

Figure 48. Structure map of the San Rafael Swell. Note that contours are structure contours.

Figure 49. West-northwest–directed photograph of the San Rafael monocline. Navajo Sandstone (white) dips very steeply southeastward. Dark flatirons at base of cliff are composed of Carmel Formation.

Part II: Deformation-band shear zones

INTRODUCTION TO DEFORMATION BANDS

Background

Superficially, deformation bands resemble veins or dikes in sandstone, cropping out as thin, tabular, planar to curviplanar, white to light gray outcrop-scale structures (Fig. 50). They commonly project in relief from the sandstone host rock. Deformation bands went essentially unnoticed as geologic structures until the late 1970s. At that time, Aydin (1978) and Aydin and Johnson (1978) described these structures in the San Rafael Desert, the Henry Mountains, and Arches National Park in the Colorado Plateau region of southeastern Utah; and Jamison (1979) and Jamison and Stearns (1982) presented the results of discovery and description of deformation bands in the Wingate Sandstone at the Colorado National Monument near Grand Junction, Col-

Figure 50. A: Photograph of deformation bands in hand specimen of Navajo Sandstone (Jurassic), Sheets Gulch area, Waterpocket Fold. Deformation bands commonly look like quartz veins, but they are not. They are semi-brittle shear zones along which offset has occurred. Photograph by Peter Kresan. B: Photograph of dike-like outcrop expression of deformation-band shear zone in Navajo Sandstone in the Cottonwood area along the East Kaibab monocline. Note person for scale on skyline to the left of main fin.

orado. Even though deformation bands look as if they are quartz veins or quartz lodes, the seminal work by Aydin (1978) and Aydin and Johnson (1978) demonstrated that they are thin, semi-brittle shear zones that accommodate fault offset. Aydin (1978) reported that individual deformation bands are typically only several millimeters or less in thickness, along which tiny offset (several millimeters or less) has been achieved (Fig. 50A). Aydin (1978) and Aydin and Johnson (1978) observed that deformation bands commonly occur in zones, ranging in thickness from 1 cm to several meters (Fig. 50B).

Jamison (1979) and Jamison and Stearns (1982) demonstrated the tectonic origin of the deformation bands they studied in Wingate Sandstone by showing how they increased in abundance with closer proximity to discrete fault zones within the Colorado National Monument. Edwards et al. (1993) later documented this same correspondence in the Mopeman Sandstone in proximity to the Lossiemouth fault in Scotland. Furthermore, as pointed out by Antonellini and Aydin (1995), rollover structures in sandstone of the Entrada Formation in the Moab region, Utah, show a direct correspondence between curvature of strata and density of deformation bands. The deformation bands accommodated the bending through a combination of layer-parallel shearing and, in areas of high curvature, layer-oblique extensional faulting.

Deformation bands and zones of deformation bands may be one of the very last outcrop-scale structures to be discovered in ordinary sedimentary rocks, yet in terms of kinematic utility they are extraordinary. They are a preferred deformation mechanism in porous granular materials (Aydin, 1978; Aydin and Johnson, 1978; Jamison, 1979; Jamison and Stearns, 1982; Antonellini et al., 1994). Individual deformation bands are almost always associated with porosity reduction and commonly are marked by grain fracturing, grain crushing, and cataclastic flow (Fig. 51) (Aydin, 1978; Aydin and Johnson, 1978, 1983). In outcrop, the deformation bands and zones of deformation bands *do not* resemble joints or faults, because deformation bands are *not* expressed as discrete fracture surfaces. As pointed out by Antonellini and Aydin (1995), the granulation and cataclasis associated with the formation of deformation bands can cause the porosity to be reduced dramatically, almost to zero along the slip plane.

Perhaps it is not surprising that tectonic denudation of porous sandstones would produce outcrop-scale structures whose properties are distinctly different from ordinary faults. Byerlee and Brace (1968) emphasized that ordinary faulting is a consequence of stick-slip behavior. They demonstrated this fact in part through results of deformation experiments. In the course of their research they discovered that certain rocks (which they called "type 2"

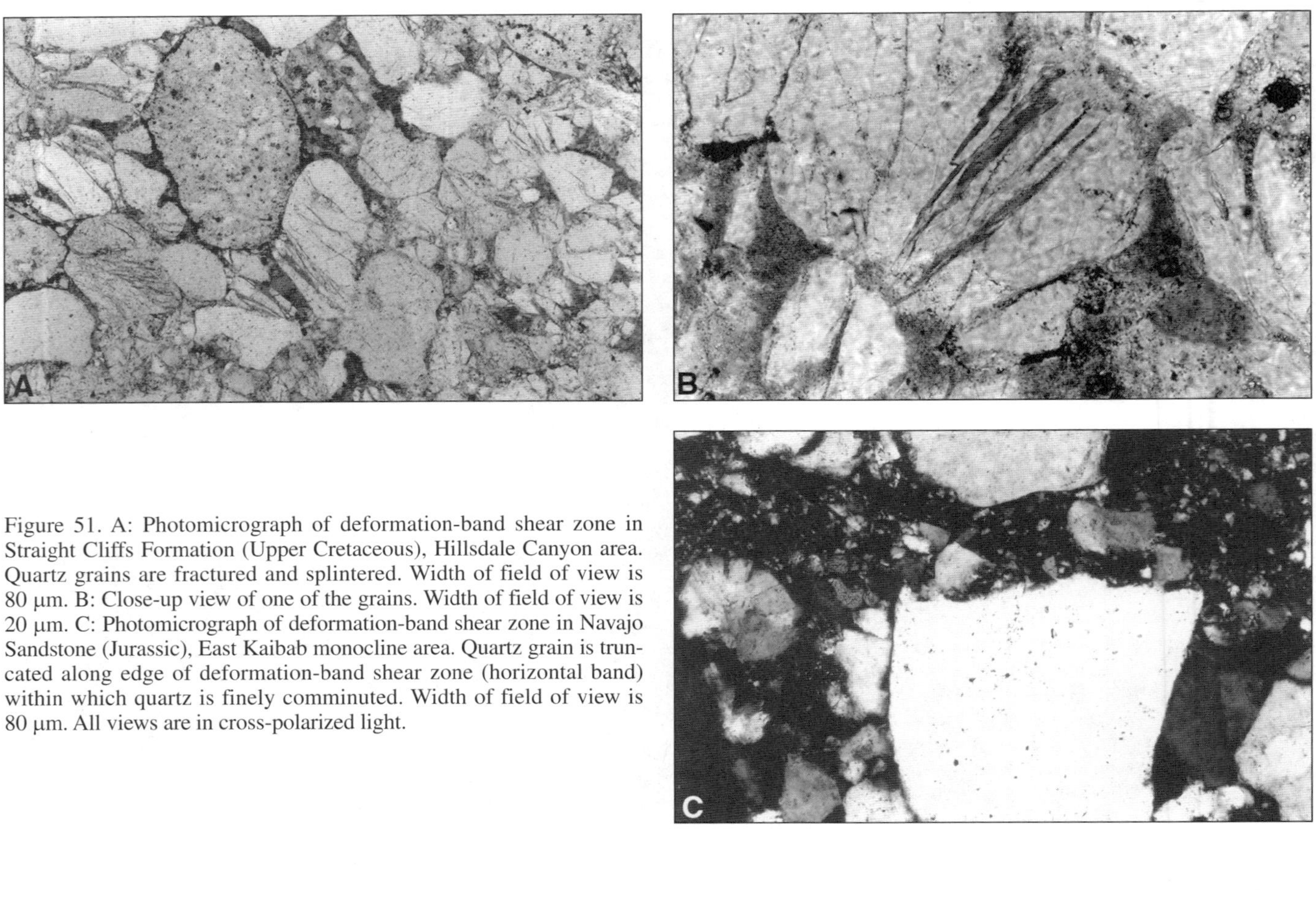

Figure 51. A: Photomicrograph of deformation-band shear zone in Straight Cliffs Formation (Upper Cretaceous), Hillsdale Canyon area. Quartz grains are fractured and splintered. Width of field of view is 80 μm. B: Close-up view of one of the grains. Width of field of view is 20 μm. C: Photomicrograph of deformation-band shear zone in Navajo Sandstone (Jurassic), East Kaibab monocline area. Quartz grain is truncated along edge of deformation-band shear zone (horizontal band) within which quartz is finely comminuted. Width of field of view is 80 μm. All views are in cross-polarized light.

rocks) do not fail via the stick-slip mechanism, but rather through a semibrittle shear involving volume loss, due to collapse of pore spaces. Among the type 2 rocks are porous sandstones!

Basic properties of deformation bands in outcrop

As described by Aydin (1978) and Aydin and Johnson (1978, 1983), individual deformation bands are about 1 mm thick and as long as 10s of meters, with offset of only a few millimeters. Shear strain within an individual deformation band ranges from about 1 to 10 (Aydin, 1978). Zones of deformation bands contain many closely spaced deformation bands across which offsets are a few centimeters or decimeters. It is not uncommon for a discrete slip surface, i.e., a fault, to form at the margin of a deformation-band shear zone. Such faults might accommodate 10s of meters of displacement.

Deformation bands and zones of deformation bands have distinctive physical and geometric characteristics in outcrop. The map-view expression of a dip-slip deformation-band shear zone, for example, is commonly marked by inosculating patterns of a braided nature (Fig. 52) (Aydin, 1978; Aydin and Johnson, 1978). (For strike-slip deformation-band shear zones, the inosculating patterns would be evident in vertical profile at right angles to the strike of the shear zone). In the plane of the shear zone, the braided habit may be expressed as an outcrop-penetrative linear structure (L-structure) that is parallel to the shear direction (Fig. 53). L-structure is parallel to slickenlines, which also occur not uncommonly on the outer margin of a zone of deformation bands. In cross-sectional view of dip-slip deformation-band shear zones, the trace expression of the shear zones tends to be quite straight.

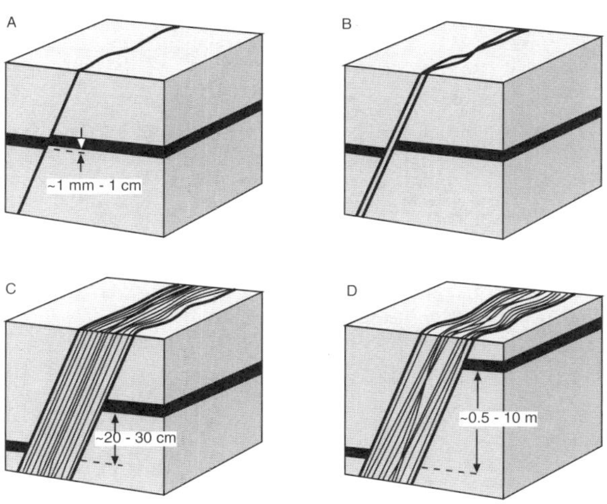

Figure 52. Schematic diagram showing the progressive development of a deformation-band shear zone. After Aydin (1978). Note how the progressive accumulation of deformation bands (A–C) leads to a braided pattern (D).

Antonellini and Aydin (1995) have emphasized that zones of deformation bands tend to occur in segments that are linked commonly in one of several ways. "Eye" structures connect two straight segments in such a way that the projections of the traces of the segments almost perfectly bisect the "eyes" (Fig. 54A).

Figure 53. A: Photograph of a deformation-band shear zone that, internally, is marked by an anastomosing, inosculating pattern of deformation bands. San Rafael Swell. Also, note lens cap (diameter, 5 cm) on top of outcrop, right center, for scale. B: Close-up view of L-structure, which here rakes ~90°. L-structure resembles ripple marks. The L-structure records direction of movement. San Rafael Swell.

Figure 54. A: Eye structure in the Entrada Formation. Sarah Tindall (left) and Jessica Graybill on slickrock. B: Ramp structure serves as connecting link at bend in deformation-band shear zone. Ramps are transpressive structures. Note relationship to sense of shear.

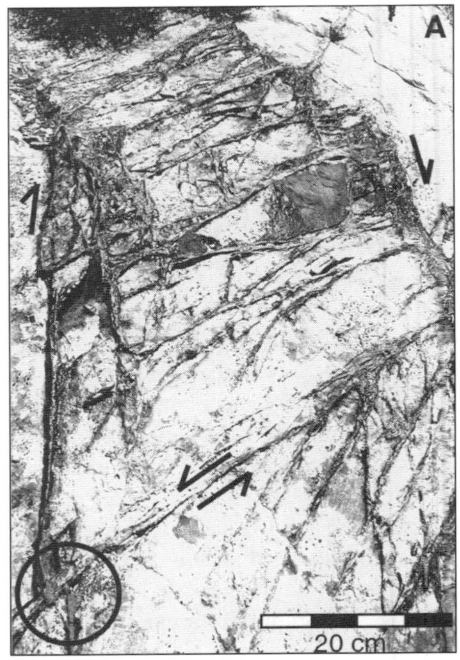

Figure 55. A: Photograph of ladder structure in an outcrop of Navajo Sandstone (Jurassic) in Capitol Reef National Park within the Sheets Gulch area. The arrangement reflects, in this case, right-handed motion of the deformation-band shear zone as a whole. The short, rung-like deformation bands are antithetic. Note the left-handed offset along the one in the immediate foreground (circled). B: Photograph of a small ladder structure in a hand specimen of Navajo Sandstone, again collected in Sheets Gulch. The arrangement, in its present orientation, reflects top-to-the-right offset. The right-dipping bands are antithetic. Photograph by Peter Kresan.

Eye structure *cannot* be used as a guide to determining sense of shear. "Ramp" structures, on the other hand, resemble miniature restraining bends (Fig. 54B), and as such they *can* be used as a guide to determining sense of shear. Within relay zones, there is a pocket of concentration of closely spaced deformation bands.

Jamison and Stearns (1982) and Antonellini and Aydin (1995) pointed out a common geometric form of deformation-band shear zones, one which I refer to as "twisted ladders" or simply "ladder structure" (Fig. 55). Ladder structure consists of two overstepping straight synthetic segments connected by antithetic "rungs." Spacing of rungs is generally related to the spacing between the overstepping segments of the major zones, i.e., the closer the spacing between overstepping synthetic segments, the closer the spacing of the rungs (Fig. 56). Orientations of the rungs relative to the ladders discloses sense of shear (Fig. 56).

Microscopic fabrics of deformation bands

Aydin (1978) discovered that deformation bands, when viewed in thin section, are seen to consist of very low porosity inner zones, within which grains are crushed to tiny angular to subangular fragments, and outer zones, within which pore spaces have collapsed (Fig. 57). Aydin (1978) and Aydin and Johnson (1978, 1983) concluded that the formation of individual deformation bands begins with a collapsing of pore spaces by a combination of deformation of the cement and rigid-body grain rotation and translation. Grains move closer and closer together, and eventually come into contact, creating stress concentrators. Internal friction will increase to the point that further deformation must involve grain-scale microfracturing. Aydin (1978) and Aydin and Johnson (1978, 1983) emphasized that angular fragments produced by the microfracturing are extremely vulnerable to continued grain-size reduction, because of the extremely high stresses that become concentrated at sharp corners. In this fashion, grains are demolished and porosity is collapsed (Fig. 58). Ultimately the deformation band "strain hardens" to the point that if continued porosity reduction and grain fracturing were to continue, these processes would take place by initiating adjacent new bands (Fig. 52). A discrete slip surface may develop, usually along the margin of the zone of deformation bands, and its presence is marked typically by polish and slickenlines. Antonellini et al. (1994) have emphasized that the deformation mechanisms and microstructures associated with the formation of deformation bands depend not only on porosity, but also on such factors as sorting, clay content, and mean stress.

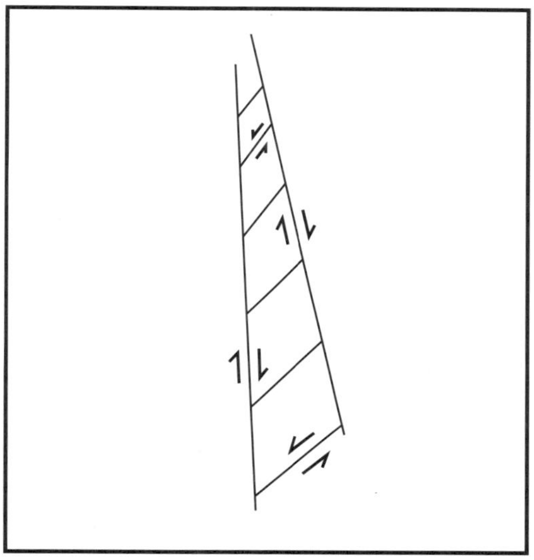

Figure 56. Ladder structure formed in the zone between two overstepping deformation bands. Note that the spacing of the rungs of the ladder is related to the distance between the overstepping deformation bands and that the orientations of the rungs relative to the poles discloses sense of shear. Taper is exaggerated.

The presence of fluids in porous sandstone may influence the development and propagation of deformation bands and deformation-band shear zones. Collapse of pore space in the process zone beyond the tip region of a deformation band may increase pore-fluid pressure directly beyond the process zone in the direction of propagation. The increased pore-fluid pressure reduces the critical shear stress required for failure. The feedback would efficiently permit the structure to propagate.

The mechanism responsible for the formation of deformation bands was first observed in association with triaxial laboratory deformation of specimens of porous sandstones. Borg (1960), for example, observed cataclastic flow along shear zones formed in St. Peters Sandstone when deformed in the laboratory. Furthermore, Friedman and Logan (1973) produced deformation bands in Coconino Sandstone that they subjected to axial loading. They called the structures "Luders' bands" and observed that these bands formed in planes of high shear stress along which cataclasis was concentrated. Dunn et al. (1973) carried out brittle-

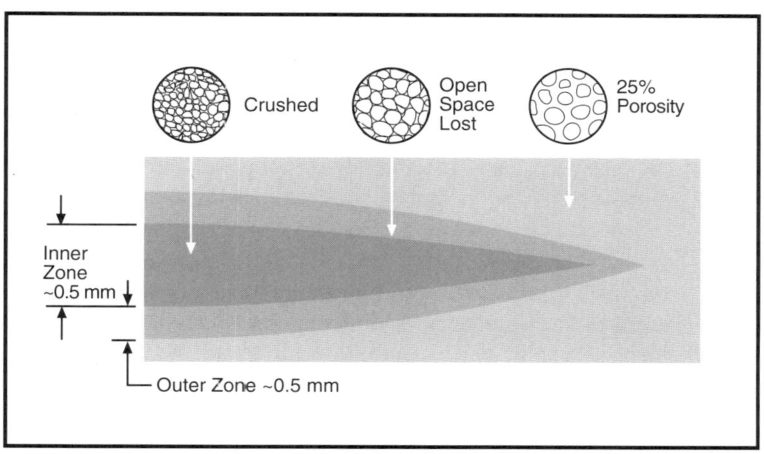

Figure 57. Schematic diagram showing the idealized anatomy of a deformation band within a highly porous sandstone host. After Aydin (1978).

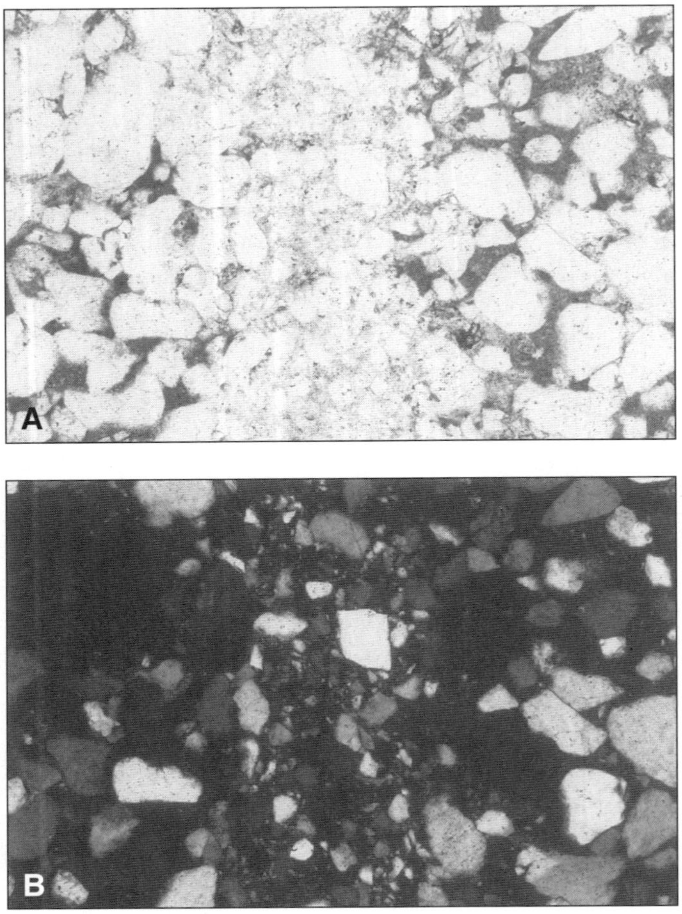

Figure 58. A: Photomicrographs of a single deformation band in Navajo Sandstone in the Sheets Gulch area. Both images are of the same area of the thin section and show the same band. Width of view of each image is 1.67 mm. The deformation band itself, which runs vertically in the center of each image, is about 0.55 mm wide. A: Open nichol. Gray areas between white quartz grains on each side of the deformation band are occupied by blue epoxy, thus giving an indication of amount of porosity. B: Cross nichol giving a clear indication of the degree of grain size reduction within the deformation band. Grain size reduction is due to cataclasis. Photomicrographs kindly provided Mingchou Lee, Mobil.

fracture tests of sandstones and found that rocks with low porosity develop throughgoing shears within which grains are reduced in size by extensional microfracturing and cataclasis. Grain-scale extensional fractures may form at stress values well below the fracture strength of the rock. Throughgoing zones of deformation are formed as a result of coalescence of grain-boundary cracks, porosity, and extension fractures.

Antonellini et al. (1994) identified two types of deformation-band fabrics. The first variety, which I call the DB1 variety, is cataclastic in nature, marked by abundant grain fracturing, modest to extreme cataclasis, and a volume reduction achieved through collapse of pore space. DB1 fabrics correspond with the deformation bands and zones of deformation bands described comprehensively by Aydin (1978) and Aydin and Johnson (1978). These typically project from outcrops in relief (Fig. 59). Resistance to erosion not only reflects the lack of porosity within these deformation bands and zones of deformation bands, but also the selective "cementing" of the fine-grained cataclastic fraction by silica. Pressure solution may be one of the sources of this silica. According to Antonellini et al. (1994), less free energy is required to nucleate the silica as the distance between the contacts decreases. The silica can even precipitate from an undersaturated solution if the distance between contacts is sufficiently small and if contacts are freshly broken. The early stage of formation of the cataclastic variety is marked by positive dilatancy, but this stage is soon overtaken by volume loss. A cataclastic origin is favored in highly porous sandstones marked by a high degree of sorting. Eolian sandstones, such as the Navajo Sandstone (Jurassic), are thus a perfect host. Antonellini et al. (1994) concluded that the grain cracking and crushing that distinguishes the cataclastic variety of deformation banding are indicative of relatively high confining pressures. In the region of Arches National Park, deformation bands of cataclastic origin typically cut and displace deformation bands of noncataclastic origin that probably formed under a very low burial load, i.e., less than several kilometers.

The presence of DB1 deformation-band fabrics has a profound influence on porosity and permeability within the porous sandstone host rock, and consequently these structures have an important bearing on the compartmentalization of reservoirs in relation to fluid flow (Antonellini and Aydin, 1994; Davis, 1997a). Antonellini and Aydin (1994) reported that the porosity within DB1 deformation bands is about one order of magnitude less than that of the host sandstone and that permeability can be reduced by several orders of magnitude. For example, Antonellini et al.(1994) measured porosities of 20% to 25% in the Moab Member of the Entrada Formation, but porosities of less than 1% to 5% in deformation bands within that same unit. Antonellini and Aydin (1994) measured permeabilities of 1,000 mD to 5,000 mD in Navajo Sandstone, which were reduced to ~2 mD to ~20 mD when measured across deformation bands.

The second variety of deformation-band fabrics (Antonellini et al., 1994), which I call DB2 fabrics, is marked by little or no cataclasis and can form with dilatancy, with compaction, or with no volume change. These form in porous sandstones under conditions of low mean normal stress. Those bands that are associated with positive dilatancy may be marked by iron oxide precipitation in pore spaces; thus DB2 fabrics are commonly are red or brown, and generally they do not weather out in positive relief.

Mollema and Antonellini (1996) have called attention to yet another variety of deformation band, which I call DB3. DB3 fabrics kinematically are equivalent to spaced cleavage. These are "compaction bands" (Mollema and Antonellini, 1996) that form perpendicular to tectonic contraction and accommodate the shortening by elimination of pore space and by loss of material through grain-contact pressure solution. Deformation bands of this variety show no evidence for shear offset. Some are pla-

Figure 59. Photograph of a resistant, vein-like expression of a deformation-band shear zone with ladder structure. Ladder structure reveals normal shear.

nar and continuous over lengths of 5–10 m; a second variety is quite crooked and very short in trace length.

Varieties of deformation bands in the study region

Two main varieties of deformation bands and fabrics were recognized in this investigation: cataclastic (DB1) and noncataclastic (DB2) (Appendix Table 1). The cataclastic deformation bands and zones of deformation bands are generally quite obvious in outcrop because (1) they are closely associated with major regional structures, such as monoclines and faults; (2) surfaces of the deformation-band shear zones commonly display slickensides and slickenlines; (3) ladder structure is abundant; and (4) grain-size reduction through grain fracturing and cataclasis is obvious. Microstructures in thin sections of the cataclastic variety range from fractured grains to splintered grains to zones of crushed, angular quartz fragments (Figs. 3, 51, and 58). Such obvious zones of cataclastic deformation are directly associated with major structures, such as faults, monoclines, and laccoliths. For example, there is usually a direct correspondence between the kinematic nature of such deformation bands and the kinematic nature of the major regional structure. Furthermore, it can generally be demonstrated that the abundance and apparent complexity of the deformation banding increase in accord with magnitude of strain and displacement along the major structure with which the bands are associated. In fact, systems of strong cataclastic deformation bands and zones of deformation bands die out where the major regional structure dies out.

Four subvarieties of cataclastic deformation bands and zones of deformation bands were distinguished on the basis of outcrop pattern (see Appendix Table 1): (1) parallel, discrete, cataclastic deformation bands and zones of deformation bands, which I call the DB1a subvariety (Fig. 60A), (2) conjugate cataclastic deformation bands and zones of deformation bands, which I call the DB1b subvariety (Fig. 60B), (3) ladder-structure cataclastic deformation bands and zones of deformation bands, which I call the DB1c subvariety (Fig. 60C), and (4) bedding-parallel deformation bands and zones of deformation bands, which I call the DB1d subvariety (Fig. 60D). All four subvarieties occur in all kinematic environments: strike slip, thrust slip, normal slip, and oblique slip.

The ladder structure (DB1c) that so often characterizes deformation-band shear zones of cataclastic origin proves to be very useful in interpreting kinematics (Fig. 61). The basic relationship is shown schematically in Figure 61A; outcrop examples are presented in Figure 61, B and C. The margins of the ladders are defined by the major deformation-band shear zones, which may accommodate thrust-, normal-, strike-, or oblique-slip movement. Within the interior of the ladder there are synthetic deformation bands aligned parallel to the margins and antithetic bands that are oblique (Fig. 61). The polarity of the acute angle between the synthetic and antithetic deformation bands gives the sense of shear for the zone as a whole. Direction of shear is perpendicular to the line of intersection of the synthetic and antithetic deformation bands, and this interpreted direction is confirmed by the orientations of slickenlines and L-structure. Ladder structures tend to occur where displacement is transferred from one deformation-band shear zone to another with which the first zone oversteps.

On the basis of outcrop character, I recognize four noncataclastic subvarieties of deformation bands (see Appendix Table 1). Almost every outcrop in the uppermost 10 to 15 m (33 to 50 ft) of the Navajo Sandstone is pervaded by the first noncataclastic variety, which I call DB2a. These consist of very thin (only as wide as several millimeters), very closely spaced, wispy, irregularly curviplanar deformation bands that occur in multiple conjugate sets and that accommodate nearly imperceptible (millimeter scale) normal displacements (Fig. 62A). Not uncommonly, individual bands can be seen to be listric, curving into parallelism with horizontal bedding (Fig. 62B). These noncataclastic deformation bands form dense networks. Most outcrops of such deformation bands reveal *multiple* conjugate normal sets, reasonably systematic in strike directions within a given outcrop, yet with such swings in strikes from outcrop to outcrop that the resultant strike rosettes cover the compass. No ladders are associated with these noncataclastic deformation bands. At the microscopic scale, these deformation bands are marked by a reduction of porosity, achieved by tighter packing of grains but without cataclasis and with minimal grain-scale fracturing.

A second distinguishable subvariety of noncataclastic deformation bands, called here DB2b, occurs in the uppermost 15 m (50 ft) of the Navajo Sandstone. This subvariety is larger than the first (DB2a) and is expressed as long, continuous, extremely well defined conjugate-normal zones of deformation bands. They are meter scale in spacing and length and as thick as 10 cm (4 in.) (Fig. 63). Offset is negligible along individual zones of deformation bands, typically on the order of several millimeters or less. No ladder structures are associated with this variety. However, quite commonly, there are restraining-bend geometries that link straight segments, and at these ramps there are lens-shaped pods where multiple deformation bands have accrued into zones. In

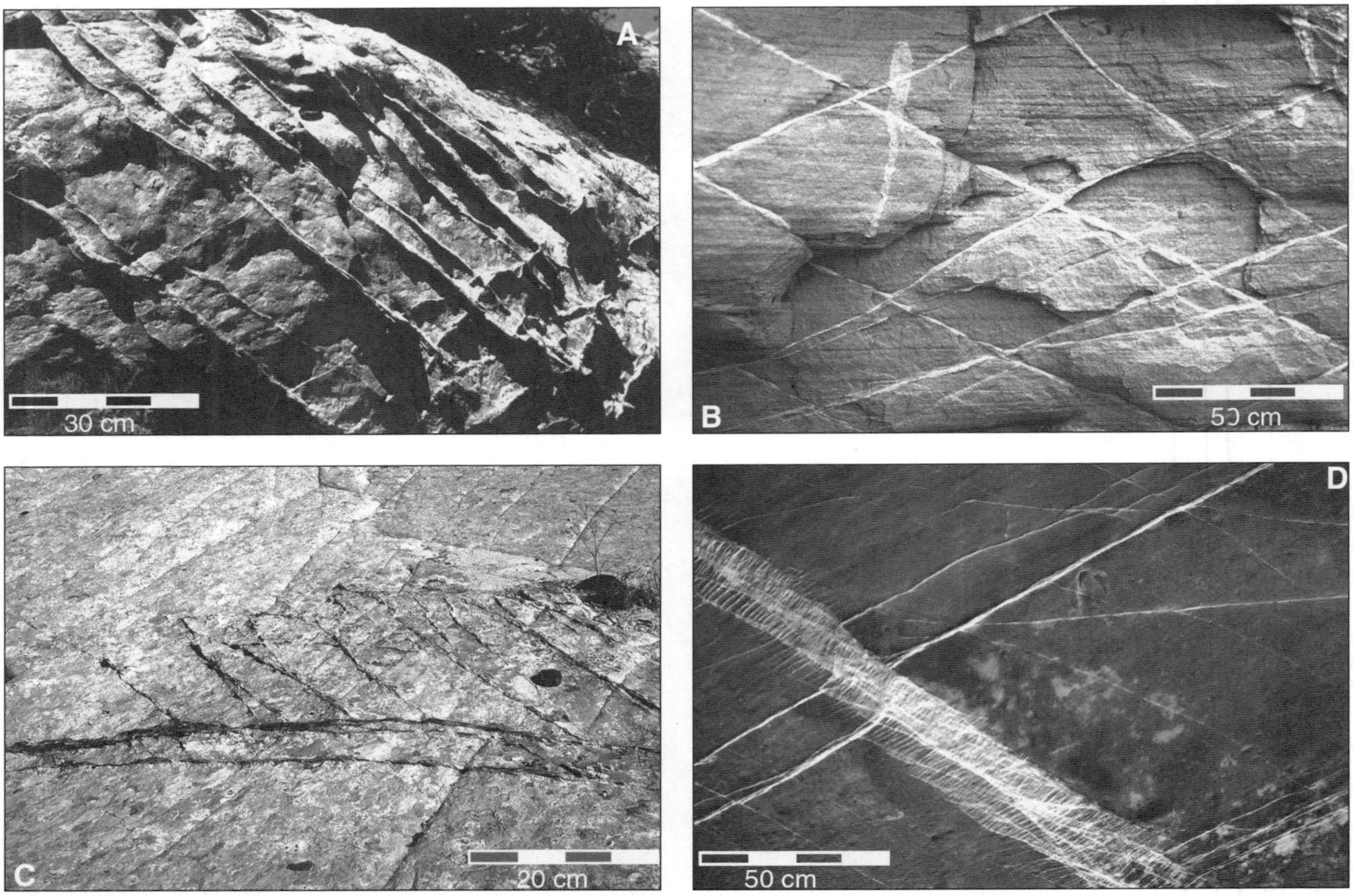

Figure 60. Four varieties of cataclastic (DB1) deformation bands and zones of deformation bands. A: Example of *parallel, discrete* deformation bands (subvariety DB1a). These normal-offset bands occur in uppermost Navajo Sandstone (Jurassic). B: Example of *conjugate* deformation bands (subvariety DB1b). These thrust-offset deformation bands occur within Straight Cliffs Sandstone (Upper Cretaceous). C: Deformation band marked by *ladder structure* (subvariety DB1c). This structure occurs in Navajo Sandstone in the Sheets Gulch area at the northernmost end of the Circle Cliffs uplift. Note lens cap. D: Some deformation bands are *bedding parallel* (subvariety DB1d), created by flexural slip along monoclines or laccoliths. Long, white layers dipping left are layer-parallel deformation bands. More delicate yet are the short, fine-scale layer-parallel deformation bands *within* the reverse-offset Riedel shear zone that dips to the right. Host rock is Navajo Sandstone. Location is in the Henry Mountains.

addition, there are braided structural patterns from place to place (Fig. 64), suggestive of linked Riedel structures (Davison, 1994, Figure 6.8a, p. 128). At the microscopic scale, it is evident that in the DB2b variety of deformation band, porosity is largely lost, yet there is minimal grain-scale fracturing.

The third and fourth subvarieties of noncataclastic deformation band (DB2c and DB2d) occur primarily in the Entrada Formation, even in places where the Entrada is essentially flat lying and far removed from regional structures (see Appendix Table 1). The DB2c subvariety consists of long, continuous, thin (as thick as several millimeters) to moderately thick (as thick as 30 cm) individual, discrete, planar to curviplanar bands. They occur as systems composed of subparallel to anastomosing and inosculating deformation bands and zones of deformation bands that are spaced at the meter scale (Fig. 65). The DB2d subvariety, which occurs almost ubiquitously in parts of the Entrada Formation, is marked by relatively closely spaced (10 to 20 cm), thin, irregular but relatively continuous deformation bands along which small fault offsets are commonly evident (Fig. 66). Seldom do these subvarieties of deformation bands in the Entrada Formation project in topographic relief. Instead, their distinctiveness is largely due to their color contrast with that of host rock: the deformation bands are either bleached white or are lighter (or darker) rose colored. Photomicrographs of the fabrics of each subvariety reveal grain fracturing, but no cataclasis. Kinematics are normal slip; individual zones dip very steeply and show offsets on the order of millimeters to perhaps as much as 30 cm (13 in.). The Entrada Formation host rock is almost everywhere crossed by these deformation bands and zones of deformation bands. Where sandstones of the Entrada Formation contain reduction spots, the reduction spots are not distorted to any extent. Deformation bands of this kind fit the description of noncataclas-

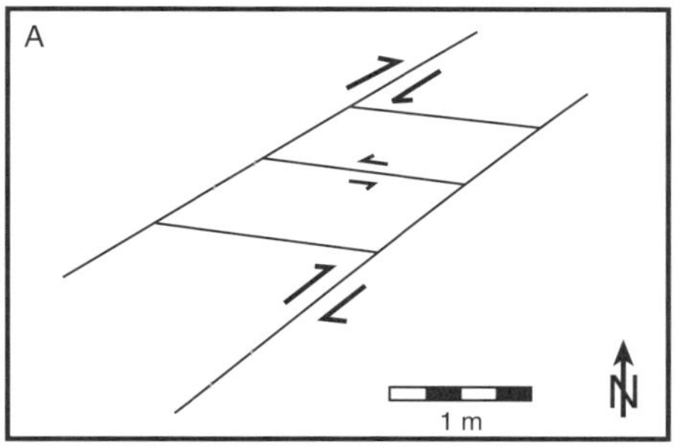

Figure 61. A: Schematic diagram showing the basic kinematics of ladder-type deformation-band shear zones. B: Photograph of boxwork ladder structure expressed in the interior of a DB1c deformation-band shear zone in the Cottonwood Canyon area along the East Kaibab monocline. Weathering and winnowing has removed porous sandstone host and has left in place the resistant DB1c deformation bands and deformation-band shear zones, creating "radiator rock." The zone dips very steeply to the left (west). Sense of shear is left side up (very high angle reverse). Clipboard in left foreground. C: Photograph of ladder structure in a float block of a weathered and eroded DB1c deformation-band shear zone. If this rock were in place, the offset would be described as top to the right.

Figure 62. Noncataclastic deformation bands of the DB2 variety. A: Photograph of closely spaced noncataclastic DB2a conjugate-normal deformation bands that occur in the uppermost 10 to 15 m (33 to 50 ft) of the Navajo Sandstone. B: Photograph showing the listric nature of some of DB2b noncataclastic deformation bands in uppermost Navajo Sandstone.

Figure 63. East-directed photograph of conjugate-normal DB2b noncataclastic deformation bands that occur commonly in the uppermost 10 to 15 m (33 to 50 ft) of the Navajo Sandstone. This locality is just south of Mount Carmel Junction.

Figure 64. Photograph of DB2b noncataclastic deformation-band shear zones near the top of the Navajo Sandstone. Inset shows a close-up of a distinctive braided pattern that occurs from place to place.

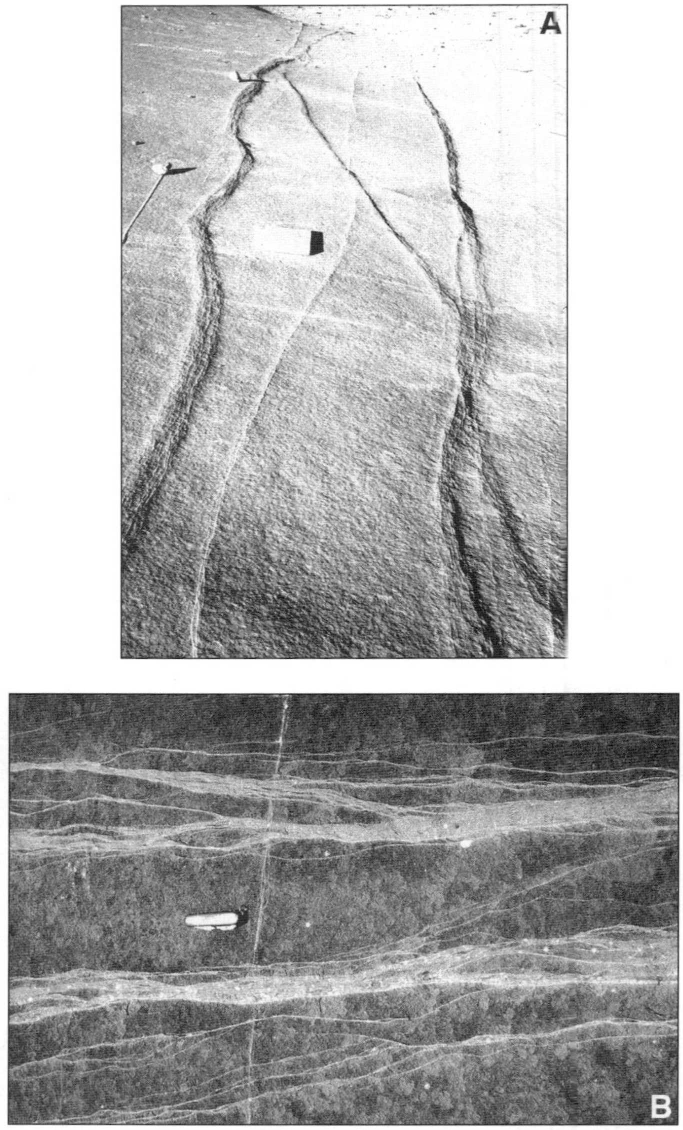

Figure 65. Photograph of outcrop expression of one of the DB2c noncataclastic subvarieties of deformation bands that occur in the Entrada Formation. A: Some examples are long, continuous, and relatively thick and tend to be smoothly braided or inosculating. Clipboard for scale. B: In places the braiding becomes interestingly complicated. Tiny white circles are reduction spots, and these for the most part show no signs of distortion. Pocketknife for scale is 10 cm long.

Figure 66. Photographs of outcrop expression of the DB2d subvariety of noncataclastic deformation bands that occurs in the Entrada Formation. A: This subvariety is irregularly continuous over broad outcrops. B: Close-up views commonly reveal offsets along these deformation bands.

tic deformation bands that form under conditions of some positive dilatancy, thus permitting hematite to move into pore space (see Antonellini et al., 1994).

Appendix Table 1 summarizes the basic properties and varieties of deformation bands and fabrics that occur in the southern Utah parks and monuments region. Appendix Table 1 includes the DB3 variety (Mollema and Antonellini, 1996), which forms perpendicular to tectonic contraction and accommodates shortening by elimination of pore space and by loss of material through grain-contact pressure solution. However, I did not recognize this variety during the course of my work, and thus it is not illustrated in Appendix Table 1. This variety occurs within the region of study, for Mollema and Antonellini (1996) described examples from the East Kaibab monocline region.

In the remainder of this chapter, occurrences of deformation-band shear zones in 78 subareas are described and interpreted for each of eight subregions: Zion, Sevier, Paunsaugunt, Kaibab, Kaiparowits, Waterpocket, San Rafael Swell, and Henry Mountains (Fig. 2A). For each subregion, Appendix Table 2 provides a subarea by subarea summary of the following information: type(s) of bands, host rock, orientation(s) of bands, orientation(s) of slickenlines, kinematic significance, inferred principal stress directions, and origin.

ANALYSIS OF DEFORMATION BANDS, ZION SUBREGION

General description of the subregion

The Zion subregion extends from the Hurricane fault on the west to the Sevier fault on the east (Fig. 67). It features some spectacular landscape and topography, including that of Zion Canyon National Park (Gregory, 1950), the Kolob region to the north and west of Zion (Fig. 68), and the headwater region of the Narrows of the North Fork of the Virgin River. Navajo Sandstone is the host rock to deformation bands and zones of deformation bands in the Zion subregion. Almost everywhere in this subregion the Navajo Sandstone dips very gently to the north-northeast; strata are essentially flat lying. Deformation bands of a noncataclastic origin (DB2) occur in the uppermost part of the Navajo Sandstone everywhere within the Zion subregion.

Outcrop properties of the deformation bands

Deformation bands and zones of deformation bands are abundant within the upper 10 to 15 m (33 to 50 ft) of the Navajo Sandstone. Uppermost Navajo Sandstone is well exposed and accessible in the Western Kolob Terrace (subarea 1), Upper Kolob Creek (subarea 2), the Kolob Terrace (subarea 3), the Zion–East Entrance (subarea 4), and the White Cliffs (subarea 5) (Fig. 67). Almost every outcrop of uppermost Navajo Sandstone contains one of two noncataclastic (DB2) subvarieties of deformation bands and zones of deformation bands: DB2a and DB2b (see Appendix Table 1). By far the most abundant is the DB2a subvariety, consisting of thin, closely spaced, somewhat curviplanar deformation bands that are essentially penetrative in outcrop (Figs. 62A and 69). The spacing typically varies from 3 to 30 cm (1 to 13 in.); trace lengths of individual deformation bands are as long as several meters, but the are commonly less than 1 m. They project resistantly and vein-like out of the bedrock, creating a sharp and hackly outcrop texture (Fig. 69). Thicknesses of these deformation bands range from just a few millimeters to several centimeters. These deformation bands are not strictly planar, but tend to be somewhat irregularly curviplanar. Typically, these deformation bands dip moderately to steeply, and from place to place it is evident that some of the deformation bands are listric in geometry, decreasing in dip as they are traced downward through the outcrop (Fig. 62B). The geometries formed by these deformation bands are conjugate normal; they show minuscule offsets on the order of just a few millimeters, although in most cases offset is not evident at the outcrop scale. Slickenlines are virtually absent on the surfaces of these bands, as are slickensided surfaces.

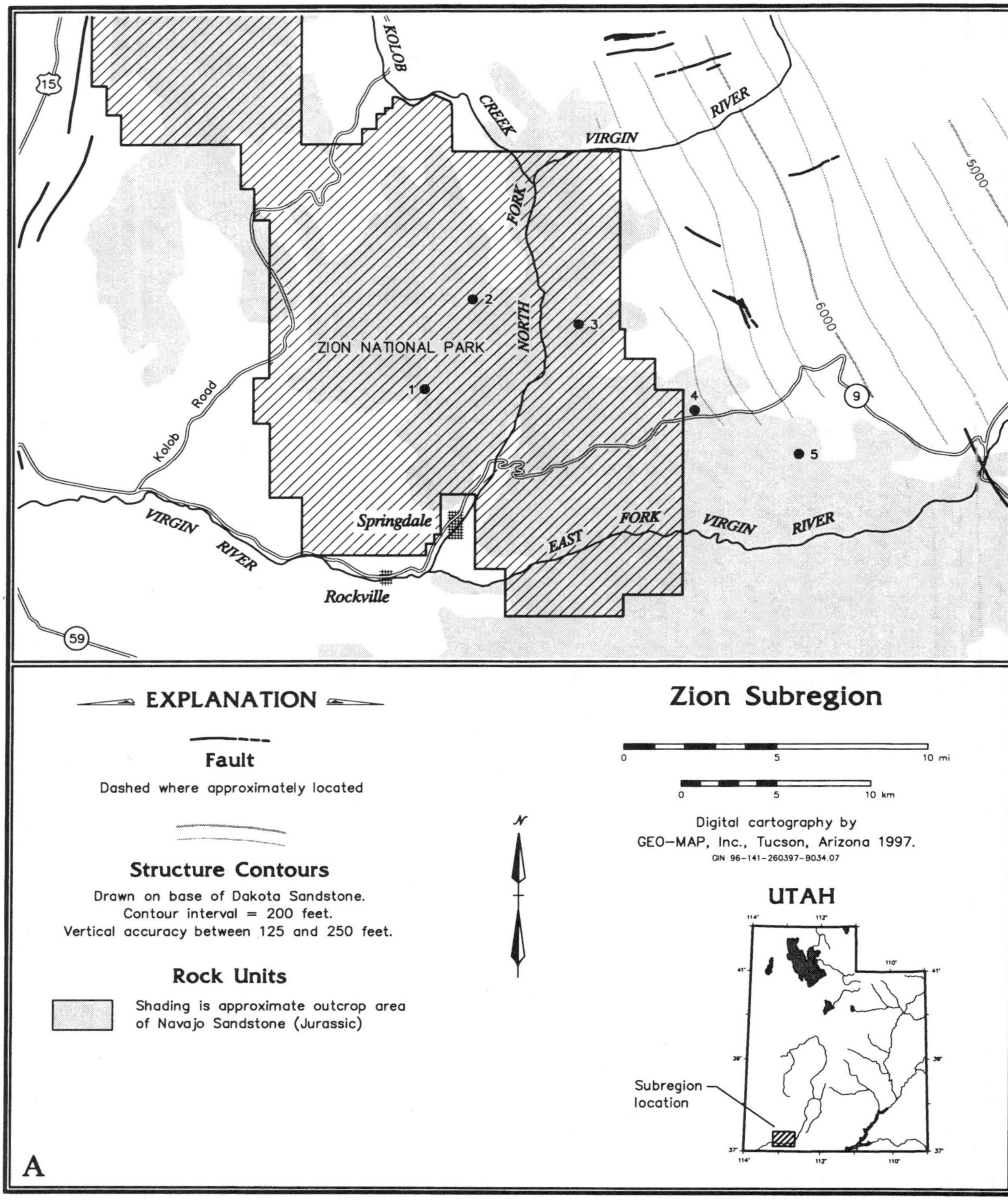

Figure 67 (on this and top of following page). A: Map of the Zion subregion, showing distribution of Navajo Sandstone and locations of subareas where deformation-band data were collected. B: Stereographic projections of deformation bands (great circles) and slickenlines (dots) measured in the Zion subregion. Numbers 1–5 refer to the subareas located in A. *N* = number of observations.

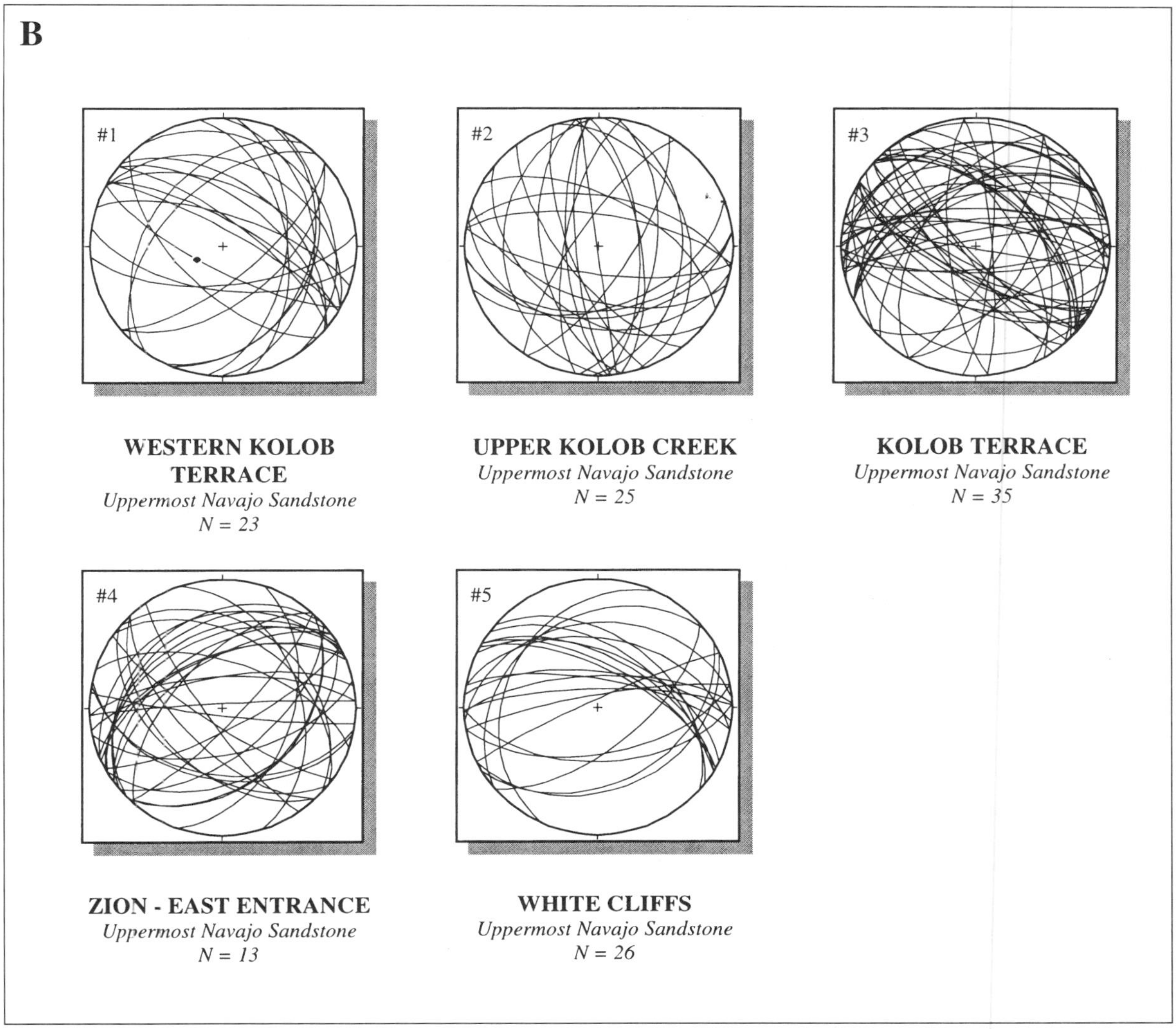

Figure 68. Photograph of Kolob Canyon in the northern reaches of Zion National Park. Top of the Navajo Sandstone (white) shows up strikingly at top of canyon. Tree-covered rolling hills at very top are underlain by Carmel Formation. Compactional deformation-band shear zones pervade the uppermost 10 to 15 m (33 to 50 ft) of the Navajo Sandstone (See arrows).

Figure 69. Photograph of the DB2a subvariety of noncataclastic conjugate-normal deformation bands that are found in the uppermost part of the Navajo Sandstone.

Perhaps the most compelling geometric characteristic of this variety of deformation band is their occurrence in multiple conjugate normal sets. No matter what the direction of cross-sectional view, an outcrop of these deformation bands will reveal conjugate normal geometries. Furthermore, in some outcrops it was possible to compare and contrast the orientations of deformation bands at one location in an outcrop with deformation bands only a few meters away in the Navajo Sandstone directly above and/or directly below. On the basis of this kind of comparative analysis, it became clear that the strike orientations of the dominant sets changed upward and downward within the sequence and laterally within the sequence as well.

Less abundant but more elegant is the DB2b subvariety of noncataclastic deformation bands. These deformation bands are also restricted to the uppermost 10 to 15 m (33 to 50 ft) of the Navajo Sandstone. They are characterized by planar to slightly curviplanar deformation bands and zones of deformation bands that are spaced several centimeters to perhaps 60 cm (24 in.) apart. Trace lengths of the spaced zones of deformation bands are as long as 8 m (26.4 ft). Thicknesses are as great as 30 cm (12 cm). They occur in conjugate-*normal* arrangements (Fig. 70). The normal displacements along this variety of deformation band are as much as several centimeters, but in many cases, the offsets are measured in millimeters. The outcrop patterns are perfect conjugate-normal structures, yet these conjugate-normal deformation bands are not systematically oriented in strike direction.

Figure 70. Photograph of the DB2b subvariety of deformation bands in the uppermost Navajo Sandstone in the Zion subregion.

Instead, strike measurements of the DB2b subvariety are found to change significantly upward, downward, and laterally within the upper 15 m (50 ft) of Navajo Sandstone.

The best and most accessible displays of the DB2a and DB2b deformation bands are near the Zion–East Entrance (subarea 4) and White Cliffs (subarea 5) (Fig. 67). Dip magnitudes range from as shallow as 20° to as steep as 80°, and in some outcrops, it is obvious that some of these zones of deformation bands are listric in form. No slickenlines or slickensides were evident in these outcrops, nor ladder structures. However, relay ramps of a normal-offset nature commonly connect deformation-band segments.

A visit to Zion Canyon National Park will confirm that the Navajo Sandstone at Zion is almost wholly devoid of deformation bands and zones of deformation bands. Yet, with the aid of binoculars, a look at the very top of the Navajo Sandstone as exposed in the great towering cliffs will reveal from place to place the presence of resistant deformation bands in the top 10 m (33 ft) or so. A drive eastward out of the park will confirm the presence of these features, as exposed, for example, behind the shops just east of the East Entrance on the north side of the road.

Microscopic properties of the deformation bands

Microscopic examination of the fabrics of these deformation bands as seen in thin section reveals that they are of the noncataclastic (DB2) variety (Fig. 71). The deformation bands are marked by closely packed grains in a network essentially devoid of porosity, in sharp contrast to the much more porous host rock. Some grain-scale fracturing is evident, but not cataclasis. The deformation mechanism appears to be one of granular flow, involving collapse of pore space and grain-surface sliding.

Stereographic geometries of the deformation bands

The characteristic stereographic signature of the DB2a and DB2b varieties of deformation bands in the Zion subregion can be seen in Figure 67B, and in particular in the plots for Western Kolob Terrace (subarea 1), Upper Kolob Creek (subarea 2), and Kolob Terrace (subarea 3). Strike orientations and dip directions of the deformation bands circumscribe the compass. Where determinations of offset could be determined, *all* were found to be normal, again corresponding to a vertical flattening. The stereographic projections reveal that dip magnitudes of the deformation bands vary from as steep as 90° to as shallow as 10°, reflecting in part a tendency for the deformation bands to be listric. On the basis of field observations, it appears that the bands tend to become listric at a stratigraphic level that corresponds with the "base" of the uppermost Navajo Sandstone zone of concentration of the noncataclastic deformation bands.

The stereographically plotted measurements for deformation bands in the uppermost Navajo Sandstone tend to form "balls of string," marked by significant variability in strike directions (Fig. 67B). This pattern is especially evident at the Zion–East Entrance (subarea 4), but also shows up at White Cliffs (subarea 5). Dip magnitudes range from gentle to steep, reflecting once again the tendency toward listric geometries.

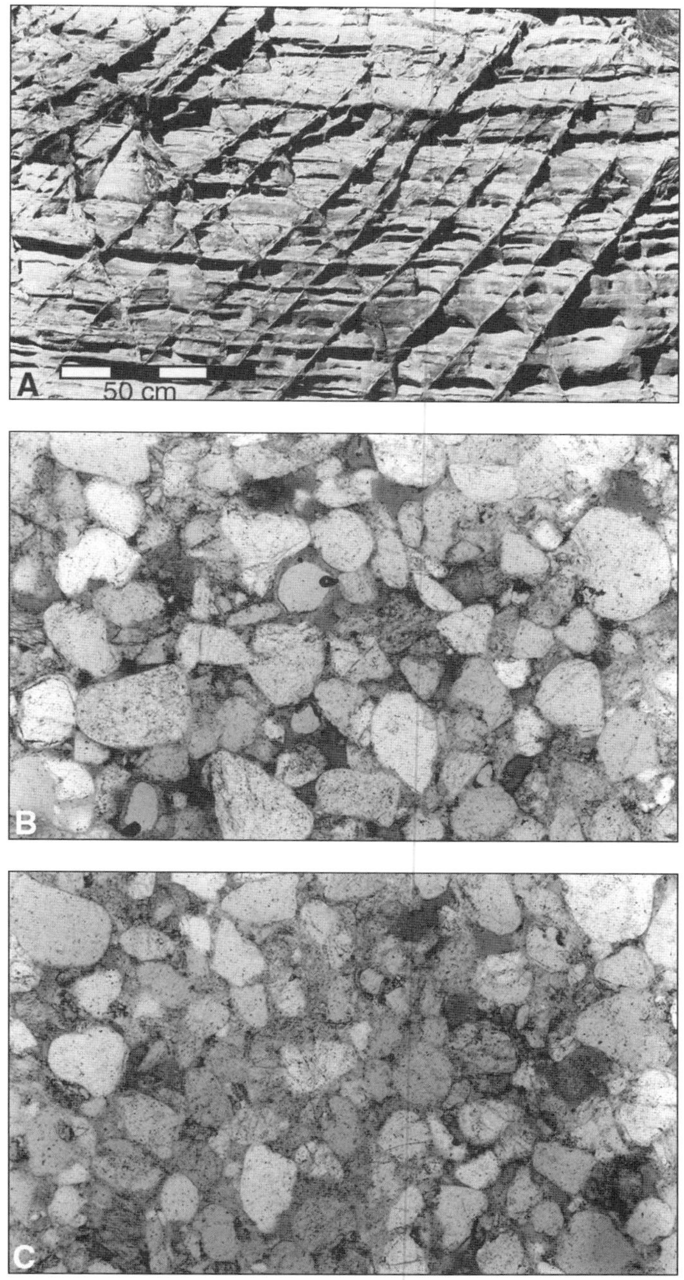

Figure 71. A: Close-up photograph of DB2a subvariety of noncataclastic deformation bands in uppermost Navajo Sandstone from which a specimen of banding was collected. B: Photomicrograph of the porous host rock in the sample collected. Width of field of view is 320 μm. C: Photomicrograph of the interior of deformation band in the sample collected. Note the degree to which porosity has been destroyed, but without cataclasis. Width of field of view is 320 μm.

Origin of the noncataclastic deformation bands

Deformation-band subvarieties DB2a and DB2b in uppermost Navajo Sandstone appear to have been produced by relatively shallow-level vertical flattening before the Navajo Sandstone was completely lithified. It turns out that this style of deformation banding in the uppermost Navajo Sandstone is pervasive across almost the entire region of study. Thus it is impossible to conclude that the vertical flattening reflected in the kinematic character of this deformation banding was accompanied by lateral horizontal extension. There is no room for accommodation. Instead, it seems more likely that the vertical flattening was a part of a volume-reduction compaction process during loading and burial of the Navajo Sandstone by the overlying Carmel Formation and, perhaps, by younger units as well. Both subvarieties DB2a and DB2b postdate soft-sediment deformation features, expressed in the form of slumps, slump folds, and liquefaction features. This structural superposition suggests that the deformation bands formed deeper and later than the formation of the slump structures, but that the overall depth of formation was less than 0.5 km (0.3 mi). In all probability, burial of the eolian Navajo sand-dune field by sediments that became Carmel Formation was accompanied by saturation of the uppermost yet-uncompacted part of the sand by ground water. Compaction may have attended a rise in the water table within the Navajo Sandstone, resulting in an increase in fluid pressure and the onset of granular flow within discrete zones. Interstitial fluid transformed the mechanical properties of the sand in ways that favored granular flow. The vertical stratigraphic interval within the noncataclastic subvarieties of deformation bands developed in the uppermost part of the Navajo Sandstone may have been related to the thickness of the water-saturated lithomechanical unit being compacted.

In short, the abundant noncataclastic deformation bands in the uppermost part of the Navajo Sandstone are *pretectonic*, formed through vertical flattening accompanying late-stage compaction during burial of the Navajo Sandstone by overlying Carmel Formation (Jurassic). In the other subregions discussed in the following sections, it becomes clear that the DB2a and DB2b subvarieties are cut and offset by deformation bands of the cataclastic (DB1) variety that most certainly are tectonic in origin.

ANALYSIS OF DEFORMATION BANDS, SEVIER SUBREGION

General description of the subregion

The Sevier subregion occupies part of the trace of the Sevier fault zone, which is an expression of Basin and Range deformation (Fig. 72) (Sargent and Philpott, 1987). The subregion is centered on Long Valley, a pastoral valley and home to several village communities that lie beneath and to the west of the cliff escarpment formed by Navajo Sandstone. The Sevier subregion projects southward through Coral Pink Sand Dunes State Park (Fig. 72). Long Valley itself follows the general trace of the Sevier fault zone. The sedimentary rocks within the subregion are Mesozoic. They strike northwest and dip very gently to the northeast (Fig. 72).

Both cataclastic (DB1) and noncataclastic (DB2) deformation bands and zones of deformation bands occur in the Sevier subregion (see Appendix Table 1). The host rock in which they occur is Navajo Sandstone. The cataclastic variety (DB1) is of tectonic origin and occurs in close association with the Basin and Range faulting expressed as the Sevier fault zone. The noncataclastic variety (DB2) occurs within uppermost Navajo Sandstone and appears to be pretectonic. Both DB2a and DB2b subvarieties of noncataclastic deformation bands are present (see Appendix Table 1).

Outcrop properties of the noncataclastic deformation bands

Good examples of the occurrence and abundance of the DB2a subvariety of noncataclastic deformation bands are found in Coral Pink Dunes State Park (subarea 6, Yellowjacket Bluff; Fig. 72). The uppermost part of the Navajo Sandstone crops out along Yellowjacket Bluff on the northwestern margin of Coral Pink Dunes State Park. Large Navajo Sandstone bedrock blocks and boulders, which appear to have been quarried in the Yellowjacket Bluff subarea, are placed at camp sites along the Loop Drive in the state park. The blocks are museum pieces of the DB2a subvariety of noncataclastic deformation bands. In actual outcrop within the Yellowjacket Bluff subarea, the DB2a deformation bands are extremely closely spaced (centimeter scale), and they pervade the outcrops (Fig. 73). The listric character of many of the deformation bands is quite apparent in single outcrops, for individual deformation bands can be traced downward from dip inclinations of 70° to dip inclinations of 20° (Fig. 73). Conjugate-normal patterns are the rule, and in most instances, offset cannot be discerned for it is so small (millimeter scale). The largest offsets recorded were on the order of only several centimeters. No slickenlines or slickensides, nor ladder structures, were observed in these outcrops. For the most part, these bands are marked by loss of pore space and some relatively minor grain-scale fracturing, though uncommonly there are very thin wispy zones of cataclastic deformation.

Another good example of the character of the DB2a subvariety of deformation bands in Navajo Sandstone is found in Red Hollow, which lies east of Glendale (Fig. 72, subarea 13). Rocks in this subarea occupy the very top of the Navajo Sandstone and display great numbers of deformation bands (Fig. 74). The deformation bands dip moderately to steeply, again with dip directions of every conceivable azimuth. Almost all of the deformation bands at the Red Hollow subarea site are oblique to the orientation of the nearby trace of the Sevier fault, which trends ~N20°E.

Examples of the DB2b subvariety of deformation bands in

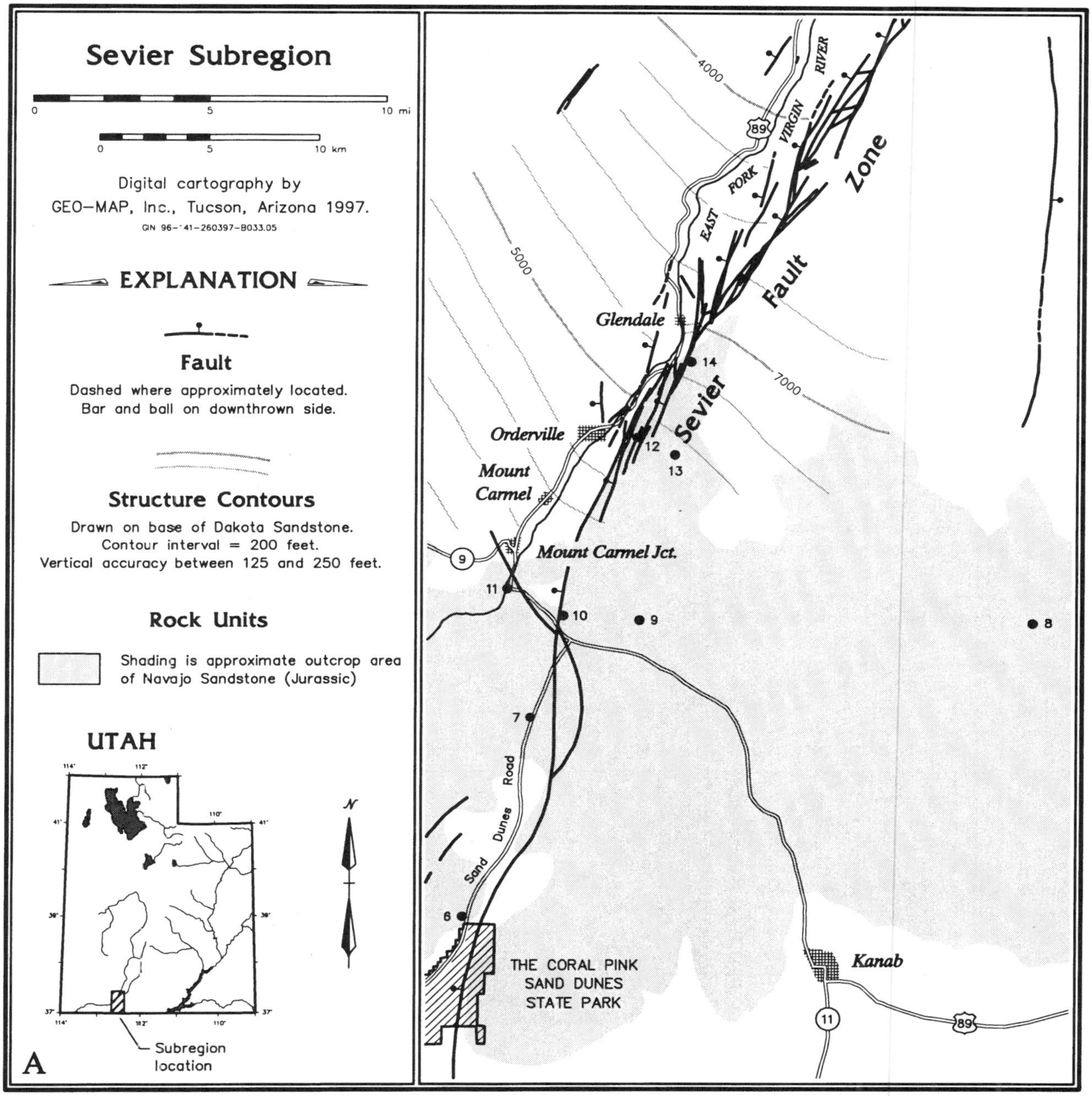

Figure 72 (this and next page). A: Map of the Sevier subregion, showing distribution of Navajo Sandstone and locations of subareas where deformation-band data were collected. B: Stereographic projections of deformation bands (great circles) and slickenlines (dots) measured in the Long Valley subregion. Numbers 6–14 refer to the subareas located in A. *N* = number of observations.

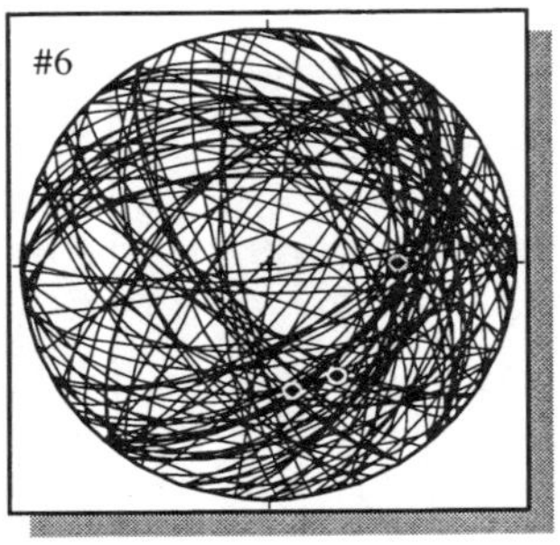

YELLOWJACKET BLUFF
Uppermost Navajo Sandstone
N = 117

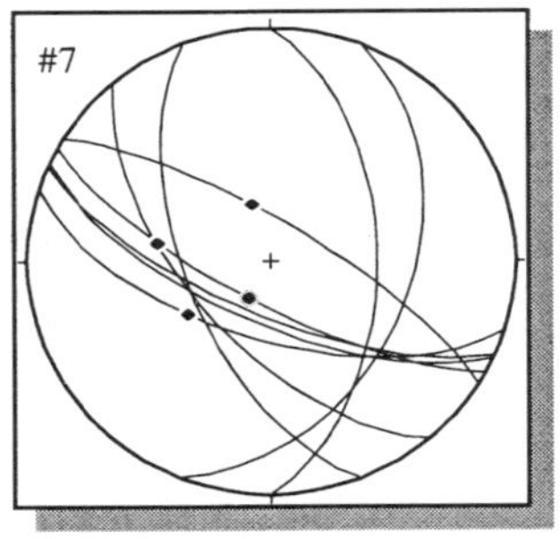

MONUMENT KNOLL
Uppermost Navajo Sandstone
N = 9

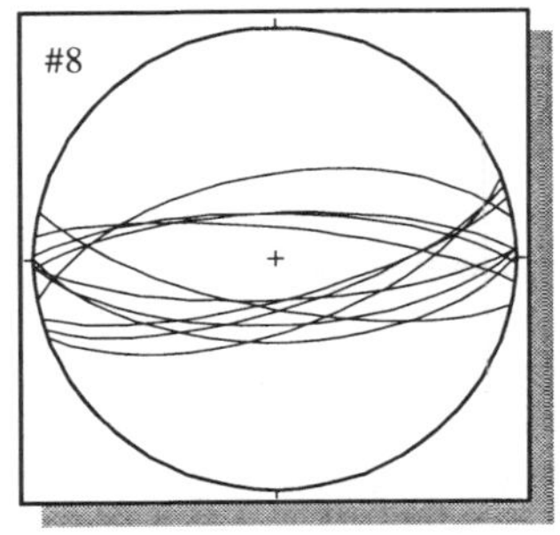

JOHNSON CANYON
Uppermost Navajo Sandstone
N = 11

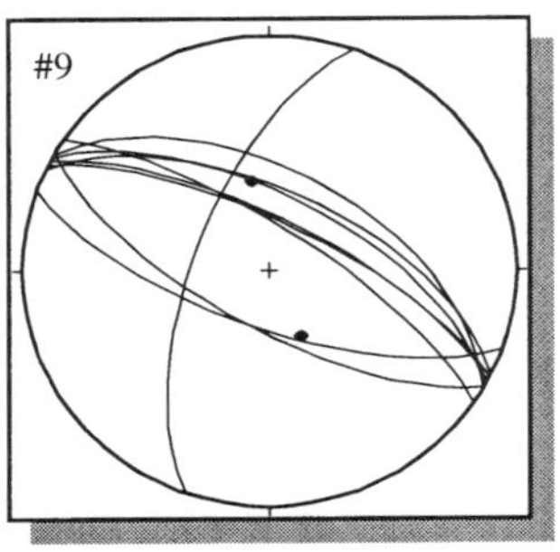

DIANA'S THRONE
Upper Navajo Sandstone
N = 9

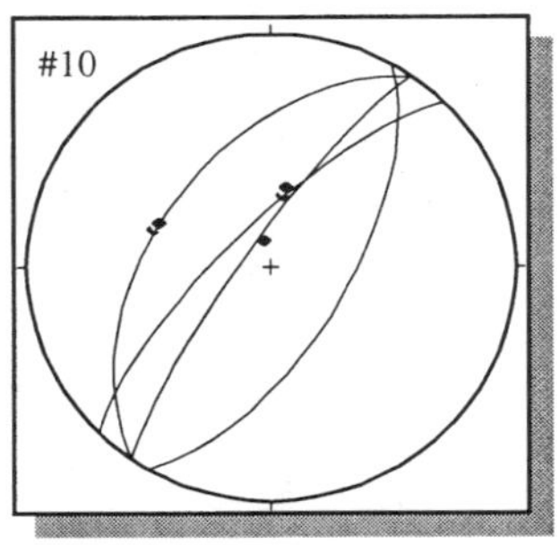

SLOT CANYON
Navajo Sandstone
N = 4

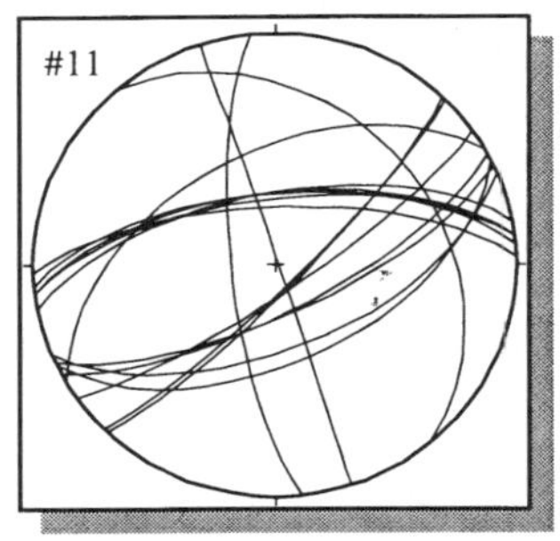

MOUNT CARMEL JUNCTION
Uppermost Navajo Sandstone
N = 16

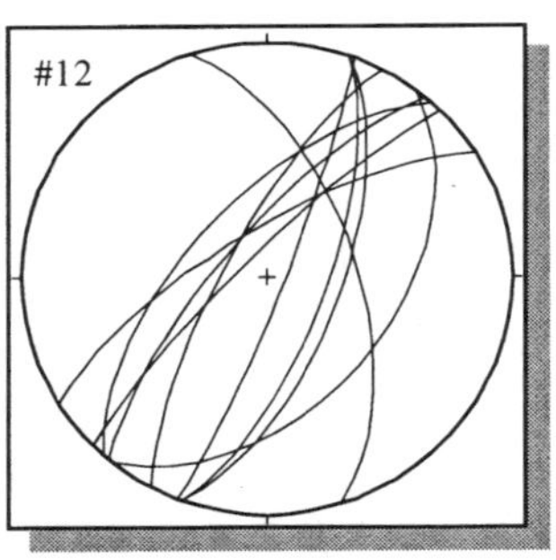

ORDERVILLE
Navajo Sandstone
N = 11

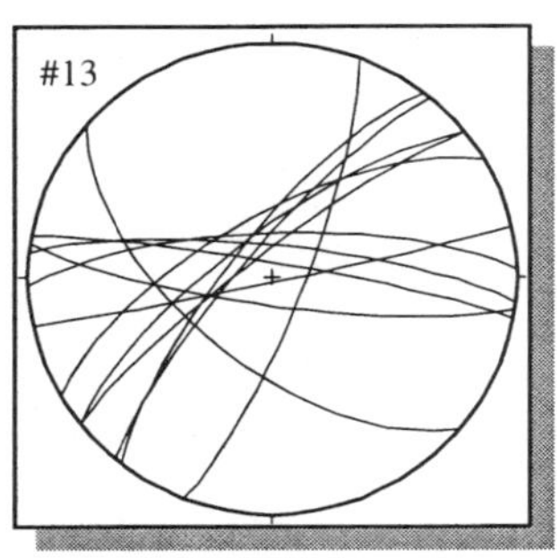

RED HOLLOW
Uppermost Navajo Sandstone
N = 12

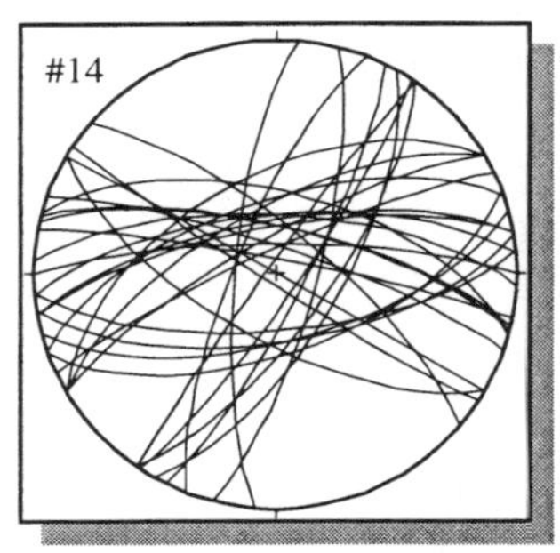

GLENDALE
Upper Navajo Sandstone
N = 32

B

Navajo Sandstone occur in a number of subareas within the Sevier subregion (Fig. 72A), including Monument Knoll (subarea 7), Johnson Canyon (subarea 8), Diana's Throne (subarea 9), and Mount Carmel Junction (subarea 11). Those at Monument Knoll, Johnson Canyon, and Diana's Throne are by no means penetrative in outcrop, but rather occur from place to place almost as isolated features that cut through the strongly cross-bedded Navajo Sandstone. Normal displacements are the rule, with offsets of as much as several centimeters. These deformation bands are as thick as 4 cm (2 in.), and the surfaces of some are marked by a steeply raking parting lineation of a noncataclastic nature (Fig. 75). Slickensides are absent.

The best examples of the DB2b subvariety of deformation bands are found at the very top of the Navajo Sandstone at a location just south of Mount Carmel Junction (Fig. 72A) on the east bank of the East Fork of the Virgin River. At this locality (subarea 11) there are magnificent exposures of conjugate-normal faults (Fig. 76). Spacing of the deformation bands ranges from several centimeters to nearly 1 m, and thicknesses of deformation bands range from <1 to 5 cm (2 in.). Toward the base of the outcrop it is evident that some of the deformation bands assume progressively lower dips, consistent with listric geometries. Offsets along the spaced deformation bands range from 1 mm to perhaps several centimeters, but in many cases, the offset is negligible. All of the offsets are normal, but there are no slickensided or slickenlined surfaces, nor are there ladder structures.

Stereographic geometry of the noncataclastic deformation bands

The widely ranging strike directions and dip directions of the DB2a subvariety of deformation bands are quite evident in stereographic projections of the measurements taken at Yellowjacket Bluff (Fig. 72B, subarea 6). The diagram looks like a "ball of string" marked by wide-ranging strike directions and significant variability in dip inclination. The stereographic projection for data collected at Red Hollow (subarea 13) also shows some of the wide-ranging strike variability that characterizes the noncataclastic deformation bands that occupy the uppermost Navajo Sandstone.

Orientations of the DB2b subvariety of deformation bands at Monument Knoll (subarea 7), Johnson Canyon (subarea 8),

Figure 73. Photograph of DB2a subvariety of noncataclastic deformation bands in the Yellowjacket Bluff subarea. Stratigraphic location is right at the top of the Navajo Sandstone. Geologist is Pilar García.

Figure 74. Photograph of DB2a subvariety of deformation bands in the Red Hollow subarea. Stratigraphic location is right at the top of the Navajo Sandstone.

Figure 75. The measuring of lineation on a DB2b deformation band exposed in Navajo Sandstone on the flank of Diana's Throne. Geologist is Ralph Colwell, though in real life he is a musician.

Figure 76. A: East-directed photograph of conjugate normal DB2b noncataclastic deformation bands in the uppermost Navajo Sandstone at the Mount Carmel subarea (11). Geologist is Kerry Caruthers. B: A closer view.

Diana's Throne (subarea 9), and Mount Carmel Junction (subarea 11) are shown stereographically in Figure 72B as well. The stereographic projections reveal that the deformation bands dip consistently from ~50° to 70° and occur in conjugate sets. With some exceptions, west-northwest is the predominant strike attitude of the DB2b deformation bands at Monument Knoll (subarea 7) and Diana's Throne (subarea 9), whereas east and east-northeast are the predominant strike directions at Johnson Canyon (subarea 8) and Mount Carmel Junction (subarea 11), respectively (Fig. 72B).

Origin of the noncataclastic deformation bands

Some of the best and most accessible outcrops of the DB2a and DB2b subvarieties of deformation bands in the Sevier subregion occur near the trace of the Sevier fault zone. Thus in the early stages of field work it was tempting to attribute the apparent extensional nature of these deformation bands to Basin and Range faulting. However, it became clear that these deformation bands are noncataclastic (Fig. 71) and unaccompanied by any of the conventional outcrop-scale indicators of tectonic origin, such as slickensides and slickenlines. Moreover, the strong N20°E preferred orientation of the strikes of fault surfaces within the Sevier fault zone does not distinguish itself in the measured orientations of the noncataclastic subvarieties of deformation bands (Fig. 72B). All of these factors suggest that these deformation bands are pretectonic, formed during Jurassic time through burial and gravitational loading of the sands now forming the Navajo Sandstone. During burial, the uppermost parts of the sand-dune field were still undercompacted. When loaded by sediments that now make up the Carmel Formation and when infiltrated by ground water, the uppermost 10 to 15 m (33 to 50 ft) failed by vertical compaction, accommodated by conjugate-normal, commonly listric, deformation bands and zones of deformation bands. The deformation mechanism was predominantly a volume-reduction granular flow. The noncataclastic subvarieties of deformation bands were thus produced.

The deformation bands are arranged in an axial symmetry about a vertical axis, supporting the interpretation of vertical compaction. Although the vertical compaction may have been accompanied locally by horizontal extension in the sand, overall the compaction involved loss of volume and was *unaccompanied* by horizontal extension.

Outcrop properties of the cataclastic deformation bands

Deformation bands and zones of deformation bands of cataclastic origin (DB1 variety) within the Sevier subregion are especially well exposed in Navajo Sandstone along the eastern margin of Long Valley, namely in Slot Canyon (subarea 10), Orderville (subarea 12), and Glendale (subarea 14) (Fig. 72).

The Slot Canyon subarea is physiographically and struc-

Figure 77. Two views of the same DB1a cataclastic deformation-band shear zone. A: North-directed view, looking along the strike of the west-dipping deformation-band shear zone. Host rock is Navajo Sandstone. B: East-directed view, looking directly at the shear-zone surface. Note the slickenlines, which rake steeply to the south (parallel to white line). Aluminum clipboard rests on right side of outcrop. Dark areas of outcrop are due to manganese staining, for the deformation-band shear zone acted as a seal to fluid flow.

turally extremely interesting. At this location, the west-dipping Sevier fault places Straight Cliffs Formation (Upper Cretaceous) on the west in direct fault contact with Navajo Sandstone (Jurassic) on the east. A west-northwest–flowing drainage cuts the fault trace at right angles, thus affording an ideal exposure of the relationships. Within the footwall Navajo Sandstone, this west-flowing drainage has carved a slot canyon, which terminates abruptly as it encounters the eastern margin of the fault zone where the canyon opens up. Quite a number of deformation-band shear zones of cataclastic origin occupy the fault zone just west of the slot canyon, where the Navajo Sandstone has been reduced to a powdery sand. North of the wash there are places where the faulted Navajo Sandstone displays zones of deformation bands in the form of fault fins (Davis, 1998), which project meters above the ground surface because of the resistance of the deformation-band shear zones to weathering and erosion. The deformation-band shear zones are of the DB1a subvariety (see Appendix Table 1). One such outcrop is 0.5 m (1.6 ft) thick, stands 4 m (13.2 ft) high, and is marked by well-developed slickenlines that rake ~75°S (Fig. 77). This zone of deformation bands is parallel to the Sevier fault zone itself, striking N10°E and dipping ~70°W. Black manganese staining of the slickenlined surface of the deformation-band shear zones records the degree to which the impermeability of such deformation-band shear-zone structures concentrates the flow of ground water.

The Orderville subarea (12 in Fig. 72) exhibits some of the best examples of deformation bands and zones of deformation bands related to the Basin and Range Sevier faulting. The Red Hollow Road, east out of Orderville, provides access to a clear view of the Sevier fault zone as exposed in the south-facing cliff exposures. The Sevier fault at this location strikes N20°E and dips 75°NW. Remarkably, right along the Red Hollow Road there is a fault-fin deformation-band exposure (DB1a subvariety) that projects more than 1 m above ground level (Fig. 78), affirming the substantial resistance of these structures to erosion. The fin exposure is fully 30 m (300 ft) south of the cliffs, projecting upward as an isolated topographic feature. The orientation of this zone of deformation bands is N20°E, 70°NW, *exactly* parallel to the overall attitude of the Sevier fault. Slickenlines rake very steeply. Ladder structure of the DB1c subvariety locally records normal-slip kinematics.

Just a short distance southwest of this location, along the Red Hollow stream bank, there is a bright white powdery exposure of faulted Navajo Sandstone. The outcrop, though quite soft and essentially cohesionless, is an exposure of bedrock replete with deformation-band shear zones of normal displacement and cataclastic character. Some of the zones of deformation bands have well-developed ladder structures (DB1c subvariety), which conform to the normal-slip displacement that one would expect at this site.

The Glendale subarea (14 in Fig. 72) also has exposures of cataclastic deformation bands relatable to the development of the Sevier fault. Navajo Sandstone in the immediate footwall of the Sevier fault zone displays abundant deformation

Figure 78. A: North-northeast–directed view. In the foreground is a DB1a cataclastic deformation-band shear zone that projects from the valley floor of Red Hollow. In the background is the trace of the Sevier fault (white line) which dips steeply west (to the left), downdropping the Navajo Sandstone. B: Closer view of the deformation-band shear zone. Chuck Kluth is the geologist.

banding of the DB1a subvariety. Zones of deformation bands are as thick as 1 m (3.3 ft). All of the offsets observed are normal; the largest offset is 16 cm (6.2 in). In places the Navajo Sandstone has been reduced by faulting to a white powdery sand. Perhaps the most striking aspect of deformation-band shear zones in this set of outcrops in the Glendale subarea is the bimodal distribution of strike orientations. The deformation bands tend to strike parallel or perpendicular to the main trace of the Sevier fault zone.

Stereographic geometry of the cataclastic deformation bands

Orientations of a few of the Slot Canyon normal-fault deformation-band shear zones (as well as slickenlines) (10 in Fig. 72B) reveal a dominant set that is essentially parallel to the Sevier fault itself, as well as a subordinate conjugate set. A stereographic plot of the orientations of the cataclastic deformation-band shear zones in the Orderville subarea (12 in Fig. 72B) reveals a system of conjugate structures striking N20°E, dipping both northwest and southeast at angles ranging from ~50° to 80°. The pattern is the same as that for Slot Canyon (subarea 10). In contrast, the orientations of deformation bands in the Glendale subarea (14 in Fig. 72) form two conjugate-normal systems, one striking parallel to the overall trend of the Sevier fault zone, the other striking perpendicular.

Origin of the cataclastic deformation bands

Within the Sevier subregion, the Navajo Sandstone appears to have escaped all deformational influence of Laramide and Miocene deformation, yet was affected by Basin and Range stresses. In particular, deformation bands and zones of deformation bands of a cataclastic and tectonic nature formed in the Navajo Sandstone within and in the immediate footwall of the Sevier fault zone. The damage zone is actually quite restricted. It is unlikely that deformation bands relatable to the Sevier high-angle normal faulting formed more than 100 m (330 ft) into the footwall.

The orientation of greatest principal stress (σ_1) responsible for formation of the Sevier fault zone was essentially vertical. The fundamental dip-slip nature of faulting suggests that the predominant regional orientation of the least principal stress (σ_3) was ~N70°W–S70°E, i.e., at right angles to the strike of the zone. The orientation of the intermediate principal stress (σ_2) was parallel to the strike of the fault, namely ~N20°E–S20°W. The actual formation of the shearing that produced the cataclastic deformation bands may have been aided by the presence of fluids in porous Navajo Sandstone Elevation of fluid pressure may have reduced the magnitude of critical shear stress required for failure. The tendency for deformation-band shear zones to form in conjugate sets both parallel and perpendicular to the trace of the Sevier fault zone (Fig. 72, Glendale, subarea 14) may reflect brief periodic exchanges of the directions of the least principal stress (σ_3) and the intermediate principal stress (σ_2). On the other hand, such bidirectional strike patterns could reflect local faulting in a three-dimensional strain field (Reches, 1978, 1983; Krantz, 1986, 1988). Three-dimensional strains could be produced within transfer zones between major fault splays of the Sevier fault zone.

ANALYSIS OF DEFORMATION BANDS, PAUNSAUGUNT SUBREGION

General description of the subregion

The Paunsaugunt subregion occupies a relatively long but rather narrow east- trending rectangular area bounded on the west by the Sevier fault zone and on the east by the Paunsaugunt fault zone (Fig. 79). The subregion contains two early Miocene thrusts of the Paunsaugunt fault zone, namely, the north-dipping Rubys Inn thrust and the south-dipping Pine Hills back thrust. It also contains the Paunsaugunt fault, which formed during Basin and Range extension (see Fig. 79). Strata within the Paunsaugunt subregion are essentially horizontal, except locally in close proximity to faulting. Fault-propagation folds are associated with the Rubys Inn thrust fault. The clearest expressions of these are in Claron Formation (Eocene) at the north end of Bryce Canyon National Park just south of Highway 12 and in Hillsdale Canyon at the western end of the trace of the Rubys Inn thrust. Also, Claron Formation strata have been dragged and tilted to dips as great as 20°W in the immediate hanging wall of the Paunsaugunt fault at the location known as "Sinking Ship" in Bryce Canyon National Park (Fig. 17).

Deformation bands and zones of deformation bands in the Paunsaugunt subregion occur in certain especially clean and well-sorted quartz sandstone units within the Straight Cliffs Formation (Upper Cretaceous). The deformation bands are cataclastic (DB1) in nature, and they occur in association with both the pre–Basin and Range, early Miocene thrust faulting and the Basin and Range normal faulting (see Appendix Table 2).

Outcrop properties of cataclastic deformation bands associated with Rubys Inn thrust faulting, Hillsdale Canyon subarea

The Hillsdale Canyon subarea (15 in Fig. 79) is an extraordinary locality, expressing the nature of conjugate-thrust deformation bands and zones of deformation bands of cataclastic origin. The structures, early Miocene in age, occur in Straight Cliffs Formation in the hanging wall of the Rubys Inn thrust. The location is near the westernmost exposure of the Rubys Inn thrust within the Paunsaugunt Plateau (Fig. 80), directly east of where the Rubys Inn thrust is truncated by the Sevier fault. A north-trending cliff exposure, ~40 m (150 ft) in height, offers a clear cross-sectional view of deformation-band shear zones produced by the thrusting. Conjugate-thrust deformation-band shear zones are abundantly exposed within the nearly flat lying sandstones in the cliff. Some of the thicker deformation bands in the cliff exposure resemble aplite dikes intruding the Upper Cretaceous sequence (Fig. 81A). A close-up view of the outcrops reveals that the "dikes" are actually resistant zones of deformation bands of cataclastic origin and with thrust displacements (Fig. 81B). The largest zones are 10s of meters in length and as thick as 1 m. The value of the outcrop lies in the penetrative nature of hundreds, if not thousands, of DB1b conjugate thrusts (Fig. 82A) as well as DB1C ladder structures (Fig. 82B). Offsets along the thrust-slip deformation bands range from millimeters to centimeters.

The resistance of the zones of deformation bands is very apparent and well expressed in interesting ways in the landscape. For example, there are a number of localities in Hillsdale Canyon where hoodoos in Straight Cliffs Formation are capped by slanted capstones in the form of thrust zones of deformation bands (Fig. 83). Some of the hoodoo caps dip northeast; others dip southwest.

The deformation bands and deformation-band shear zones in the Hillsdale Canyon area formed preferentially in porous sandstone units within the Straight Cliffs Formation, *not* in the low-porosity calcite-cemented sandstones. Where the Straight Cliffs Sandstone is composed of calcite-cemented sandstone, slickenlined fault surfaces formed instead. It is especially interesting to observe the two contrasting lithologies in a single outcrop and to see the association of slickenlined surfaces confined to calcite-cemented sandstones and the deformation bands restricted to the more porous sandstones.

Some of the deformation bands and zones of deformation bands in the Hillsdale Canyon area affect granule conglomerates that occur as lenses within the sandstone. From place to place, pebbles are highly fractured and offset by faulting along deformation-band shear zones, and the sense of offset based on displaced pebbles constitutes some of the best documentation of the thrust nature of the structures at this locality. Figure 84 is a photograph of a slab of sandstone and granule conglomerate that is cut by deformation bands. The surface of this specimen (Fig. 84A) reveals white, frosty, regularly oriented traces of sets of deformation bands in the sand-rich parts of the specimen. One set is bedding parallel (DB1d subvariety). The other set occurs at a high angle to bedding, making an angle of ~75° to the first. The sawed and polished surface of this slab (Fig. 84B) reveals interesting structural details. For example, some of the bedding-parallel DB1d subvariety of deformation bands are located exactly along the contact between sandstone and conglomerate. Furthermore, the traces of deformation bands in sandstone extend across conglomerate layers, within which granules and small pebbles are offset in stair-step fashion.

The deformation-band geometries in the Hillsdale subarea exhibit pervasive DB1c ladder structure (Fig. 82B), consisting of long, continuous, synthetic thrust-slip deformation-band shear zones connected by short, discontinuous, antithetic thrust-slip deformation-band shear zones. Offsets of both original bedding and deformation bands are obvious and abundant, making it straightforward to distinguish "synthetic" vs. "antithetic" deformation bands within a given outcrop area. The ladder structures appear to be best developed where thrust-slip deformation-band shear zones overstep one another in the updip or downdip directions. It is within the areas of overstepping that the ladder structures occur. The cliff exposure at Hillsdale Canyon offers many examples of such ladder structures, and it becomes immediately

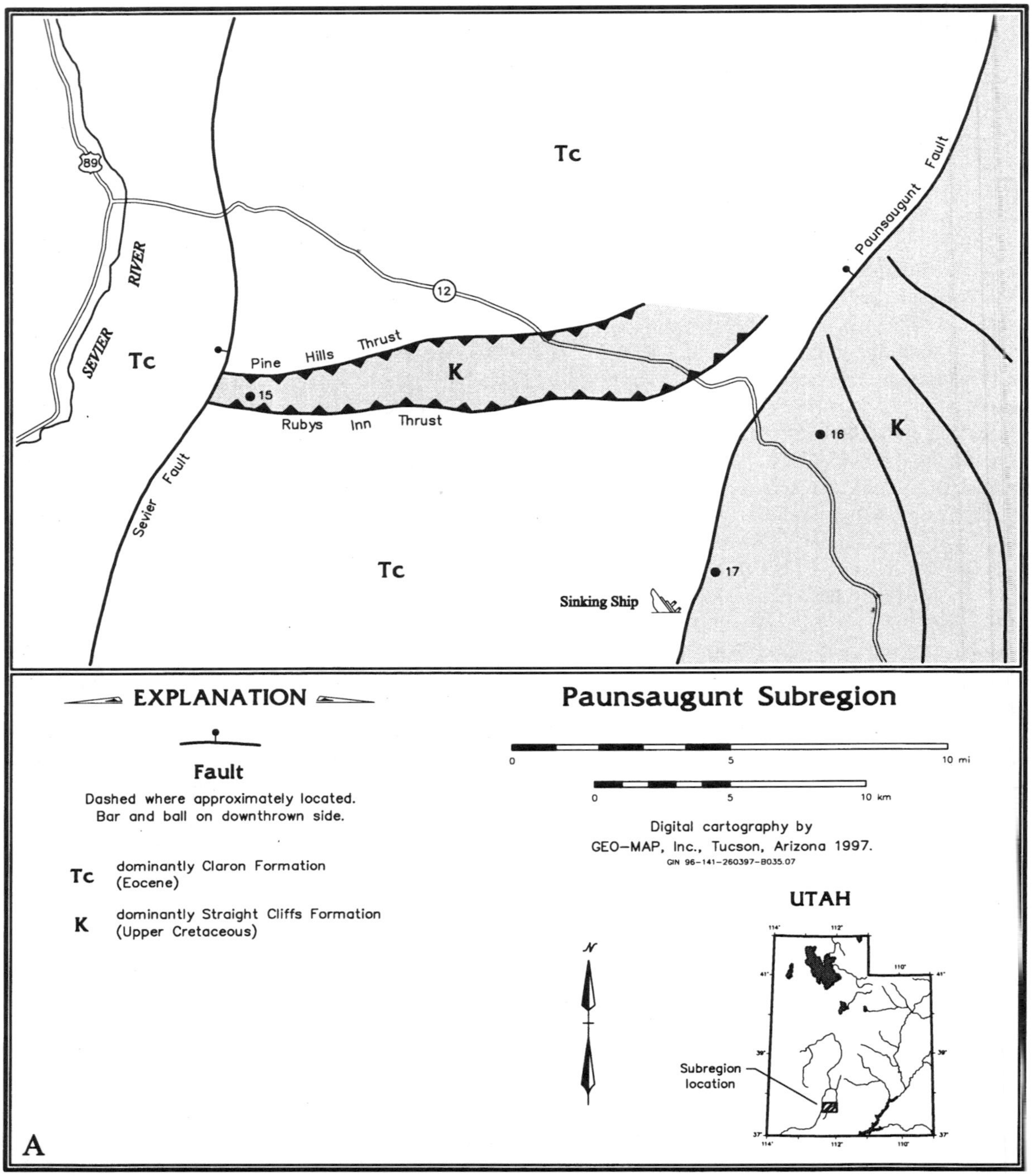

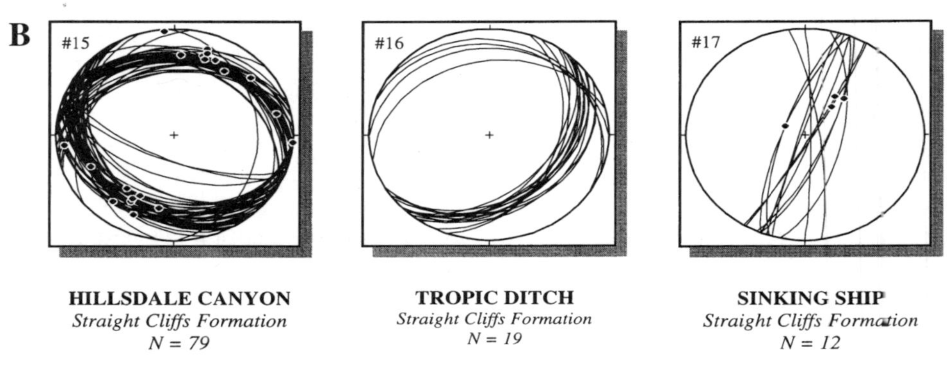

Figure 79. A: Map of the Paunsaugunt subregion, showing locations of subareas where deformation-band data were collected. B: Stereographic projections of deformation bands (great circles) and slickenlines (dots) measured in the Paunsaugunt subregion. Numbers 15–17 refer to the subareas located in A. N = number of observations.

Figure 80. The Hillsdale Canyon area lies along the trace of the Rubys Inn thrust, whose position here is shown in an east-directed photo. Ksc—Straight Cliffs Formation (Cretaceous), Tc—Claron Formation (Eocene).

Figure 81. A: West-directed view of cliff of Straight Cliffs Formation, showing long, continuous deformation-band shear zone that dips northward. Bedding is essentially horizontal. B: Closer view of this major deformation-band shear zone, along with multiple south-dipping deformation-band shear zones in its footwall. The geologist is Alex Bump.

apparent that the spacing of antithetic structures within the relay zones bears a direct relationship to the spacing of the overstepping synthetic segments (Fig. 82B).

The thickest zones of deformation bands dip north at some locations in the Hillsdale Canyon subarea, but they dip south at other locations. In all probability there is a "balancing" of dip directions in ways that minimized the overall rotation of the Straight Cliffs Formation bedding during shortening. Indeed, the bedding is essentially horizontal, and the deformation viewed as a whole is essentially a coaxial "pure" shear. Some of the deformation banding may well have developed just prior to the propagation of the tip of the Rubys Inn thrust through

Figure 82. A: West-directed view of outcrop of near-perfect DB1b conjugate-thrust deformation-band shear zones. B: West-directed view of outcrop of DB1c ladder-structure deformation-band shear zones, which are very abundant in the Hillsdale Canyon subarea. Host rock is Straight Cliffs Formation (Upper Cretaceous).

Figure 83. East-northeast–directed photograph of hoodoo landscape in the Hillsdale Canyon subarea. The hoodoos are capped by gently inclined, resistant, rock slabs, that are remnants of cataclastic deformation-band shear zones.

the stratigraphic levels of exposure revealed at the Hillsdale Canyon locality. If so, the conjugate-thrust deformation-band shear zones would be expected to have remained symmetrical with respect to bedding, even where Straight Cliffs Formation is tilted to nearly vertical in the footwall syncline of the Rubys Inn thrust.

Outcrop properties of cataclastic deformation bands associated with Rubys Inn thrust faulting, Tropic Ditch subarea

There are other locations in the Paunsaugunt subregion where conjugate-thrust deformation bands related to the Miocene thrusting may be observed, though none rivals the structures at Hillsdale

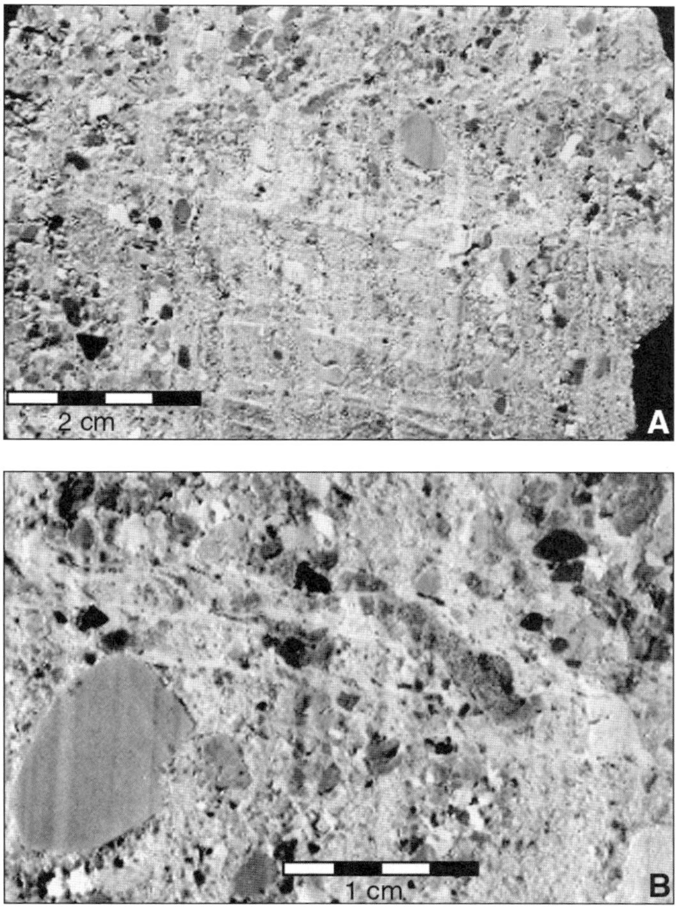

Figure 84. Photographs of a slab of sandstone and granule conglomerate collected by Chuck Kluth (Chevron, Inc.) in the Hillsdale Canyon area. A: The white frosted traces are deformation bands. B: Close up view of faulted pebbles. Clean faults cut through pebbles; deformation bands cut through the sand matrix. Photographs by Peter Kresan.

Canyon. But one location that is worth mentioning is the Tropic Ditch subarea (16 in Fig. 79), east of the Paunsaugunt Basin and Range normal fault. The host rock is Straight Cliffs Sandstone. The deformation bands at the Tropic Ditch locality are of the cataclastic DB1b subvariety, i.e., conjugate-thrust structures, with very small thrust offsets ranging from millimeters to a few centimeters. The deformation bands tend to be closely spaced (10 to 30 cm [4 to 12 in.]) and occur in lenses and layers of favorable lithology. These structures appear to have been formed in the footwall of the Rubys Inn thrust, before the onset of Basin and Range faulting along the Paunsaugunt fault, which lies just 2.5 km (1.5 mi) to the west.

Stereographic geometry of the early Miocene deformation bands

The orientations of the zones of deformation bands, as revealed stereographically, are among the most systematic measured in this entire study (Fig. 79B). In the Hillsdale Canyon subarea (15 in Fig. 79B), they strike N65°–75°W and dip from 20° to 35° both to the northeast and to the southwest. Slickenlines trend, on average, S15°W. The slickenlines are uncommon but, where observed, reveal a direction of tectonic transport that is parallel to the direction of transport on along the closest exposures of the Rubys Inn thrust. Direction of transport is essentially perpendicular to the line of intersection of the conjugate thrusts, i.e., S15°W (Fig. 79B)

The orientations of conjugate-thrust deformation-band shear zones in the Tropic Ditch subarea (16 in Fig. 79B) are very systematic as well, striking N50°E and dipping moderately, mainly to the southeast but also to the northwest.

Cataclastic deformation bands associated with the Paunsaugunt faulting

Within the Paunsaugunt subregion there are also conjugate DB1b *normal-slip* deformation-band shear zones of cataclastic origin, and these occur in the Sinking Ship subarea (17 in Fig. 79A). They occur in Straight Cliffs Formation in the immediate footwall of the Paunsaugunt fault. The outcrops are riddled with small-scale normal-slip deformation bands, reflecting the cataclastic deformation mechanism that operated in the fault zone during the extensional deformation. Stereographic projection of the measurements of deformation bands at the Sinking Ship reveal that the deformation bands strike parallel to the strike of the Paunsaugunt fault (i.e., N10°E) (17 in Fig. 79B). Dips are steep, ~70°, both to the southeast and to the northwest. Slickenlines are moderately to steeply raking.

Origin of the deformation bands

The cataclastic deformation bands in the Hillsdale Canyon and Tropic Ditch subareas (Fig. 79A) resulted from horizontal shortening related to the development of the early Miocene Paunsaugunt thrust system. There are several lines of evidence to support this interpretation. First, the systems of deformation-band shear zones are well developed only in close or relatively close proximity to the trace of the Rubys Inn thrust. Second, in both the Tropic Ditch and the Hillsdale Canyon subareas, the direction of shortening interpreted from the orientations of deformation-band shear zones was parallel to the direction of slip determined from slickenlines on the Rubys Inn thrust (and associated structures). In the Tropic Ditch subarea, the direction of shortening is ~N40°W–S40°E, and the direction of thrusting was ~S40°E. In the Hillsdale Canyon subarea, the direction of shortening based on deformation bands was ~N30°E–S30°W, and the direction of thrusting was ~S30°W. Thus, at any one location along the thrust belt, there was a close spatial association between thrust faulting and the development of conjugate-thrust deformation bands, and there was a close kinematic coordination between the direction of shortening revealed by the systems of conjugate-thrust deformation bands and the direction of movement along the Rubys Inn thrust. This statement is true even though the Paunsaugunt thrust belt itself is convex southward in shape, curving from southwest to a west trend as the belt is followed across the Paunsaugunt Plateau from east to west (Fig. 25). The direction of sense of slip of thrusts within the belt changes from southeast to southwest as the belt is followed across the Paunsaugunt Plateau from east to west (Fig. 25).

Directions of principal stress can be inferred from the orientations of deformation bands for the Hillsdale Canyon and Tropic Ditch subareas. For the Hillsdale Canyon subarea, the greatest principal stress (σ_1) was horizontal and oriented ~N30°E–S30°W, the intermediate principal stress (σ_2) was horizontal and oriented N60° W–S60° E, and the least principal stress (σ_3) was essentially vertical. For the Tropic Ditch subarea, the greatest principal stress (σ_1) was horizontal and oriented N40°W–S40°E, the intermediate principal stress (σ_2) was horizontal and oriented N50°E–S50°W, and the least principal stress (σ_3) was essentially vertical.

The cataclastic deformation bands in the Sinking Ship subarea (17 in Fig. 79B) resulted from horizontal extension related to the development of the Basin and Range Paunsaugunt fault. The deformation bands are found only in the immediate footwall of the fault and only where lithologies of footwall rock are favorable. For the Sinking Ship subarea, the greatest principal stress (σ_1) was vertical, the intermediate principal stress (σ_2) was horizontal and oriented N10°E–S10°W and the least principal stress (σ_3) was horizontal and oriented N80°W–S80°E.

With respect to deformation bands related to Basin and Range faulting, observations in the Paunsaugunt subregion corroborate what was learned in the Sevier subregion. Deformation bands and zones of deformation bands related to the extensional Basin and Range faulting always seem to lie in very close proximity to the trace of such faults. On the other hand, deformation bands and zones of deformation bands related to the compressional early Miocene deformation can form significant distances away from the trace of the major thrusts, as originally pointed out by Nickelsen et al. (1992). Nickelsen et al. (1992) mapped "deformation fronts" relatable to the Rubys Inn thrusting (shaded bands in Fig. 25). Each deformation front defines the southernmost limit of a specific outcrop-scale feature produced during Rubys Inn thrusting. For example, the closest "front" to the trace of the Rubys Inn thrust represents the limit of slickenlined shear surfaces. These occur in the footwall no farther than 5 km (3 mi) south of the fault trace. The deformation front representing the development of deformation-band shearing related to early Miocene compression lies ~10 km (6 mi) south of the trace of the Rubys Inn thrust (Fig. 25).

ANALYSIS OF DEFORMATION BANDS, KAIBAB SUBREGION

General description of the subregion

The Kaibab subregion extends southward from Kodachrome Basin State Park, located southeast of Tropic, to the Utah-Arizona state line (Fig. 85). The main access is the Cottonwood Canyon Road, a spectacular drive along strike valleys in the upturned Mesozoic section of the East Kaibab monocline (Figs. 35B and 37A). The Cottonwood Canyon Road connects Utah Scenic Highway 12 with Highway 89 at a location near the trailhead for hiking the Paria Canyon Narrows. The Kaibab subregion also includes the crestal region of the Kaibab uplift, to the west of the monocline, as well as the westernmost edge of the Kaiparowits basin, to the east of the monocline. Navajo Sandstone occupies the crestal region of the Kaibab uplift throughout much of its length in Utah. The structural transition to the Kaiparowits basin to the east is marked by well-exposed monoclinal folding of Upper Cretaceous Straight Cliffs Formation and Wahweap Sandstone (Fig. 86). The Kaibab uplift dies out to the north in the form of a broad, gently plunging anticline. The north plunge of the structure is evident in the vicinity of Kodachrome Basin State Park (Fig. 2, B and D), where Jurassic strata of the Carmel and Entrada Formations dip gently north (no more than ~5°).

Deformation bands and zones of deformation bands in the Kaibab subregion abound in the Navajo Sandstone and in parts of the Straight Cliffs Formation as well. The densely clustered noncataclastic DB2a subvariety of deformation bands everywhere occupies the uppermost part (top 15 m [50 ft]) of the Navajo Sandstone, and these probably formed during burial and compaction of the Navajo Sandstone. These deformation bands are also evident in places in the Page Sandstone Tongue of the Carmel Formation (Jurassic). In addition, some units within the Straight Cliffs Formation and Wahweap Sandstone (Upper Cretaceous) are pervaded by cataclastic (DB1) deformation bands and zones of cataclastic deformation bands. These occur in geometric arrays that are among the most structurally complex in the entire region of study (see Appendix Table 2).

Properties of noncataclastic deformation bands in Navajo and Page sandstones

Noncataclastic deformation bands of the DB2a subvariety are conspicuous in the Kaibab subregion within the Page Sandstone Tongue and the uppermost Navajo Formation where they mainly occur; they are very well exposed and quite accessible along most of the full length of the Cottonwood Road. Thus, some of the first structures that one encounters hiking westward from Cottonwood Road into the monocline are the systems of DB2a deformation bands. Before it became evident that these structures were pretectonic, it was natural to assume that these deformation bands developed during the Laramide, accompanying the formation of the East Kaibab monocline.

There are several subareas where the DB2a subvariety of deformation bands is especially well exposed. For example, the Lower Slickrock subarea (18 in Fig. 85B) contains very gently dipping white sandstone of the Page Sandstone Tongue of the Carmel Formation on the north plunge end of the Kaibab uplift. Outcrops are full of deformation bands that conform to the DB2a subvariety. They are thin, relatively short, and closely spaced.

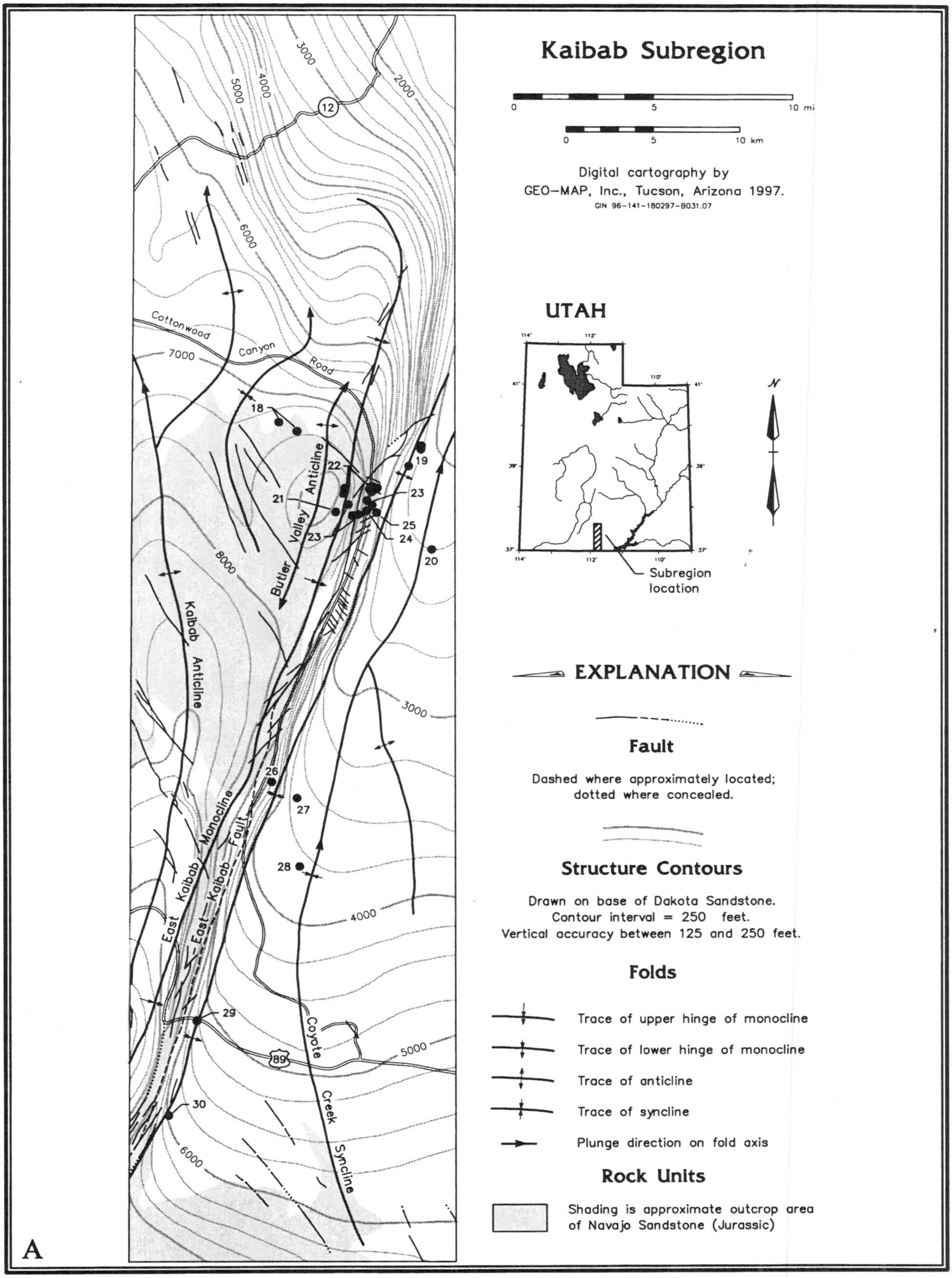

Figure 85 (this and following two pages). A: Map of the Kaibab subregion, showing distribution of Navajo Sandstone and locations of subareas where deformation-band data were collected.

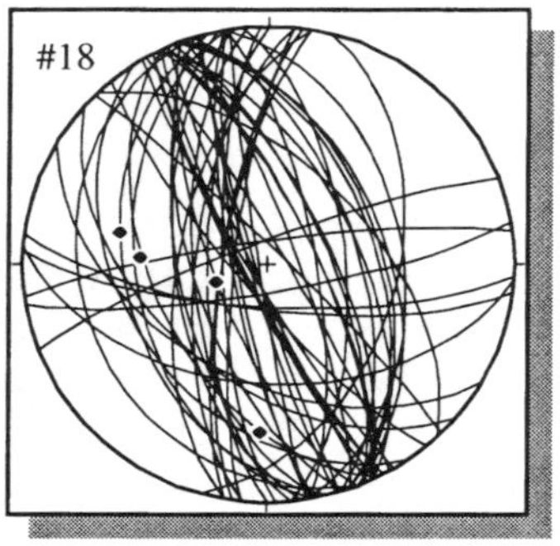

LOWER SLICKROCK
Uppermost Navajo Sandstone
N = 54

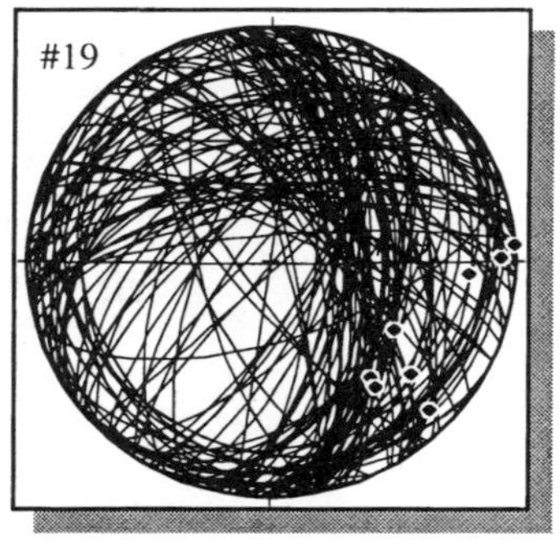

CAD'S CROTCH
(Undifferentiated)
Straight Cliffs Formation
N = 150

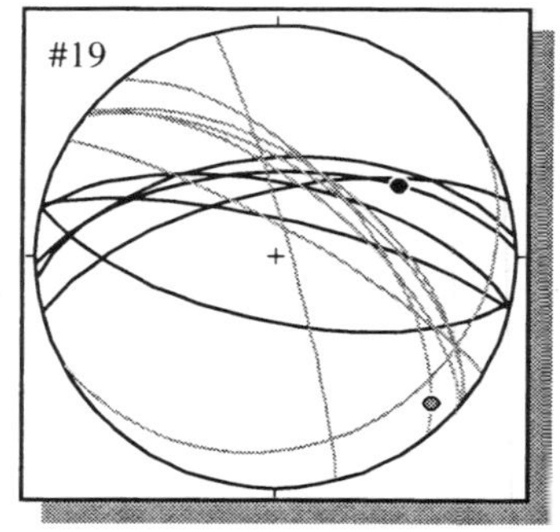

CAD'S CROTCH
(Right and Left Lateral)
Straight Cliffs Formation
RL = 6   LL = 7

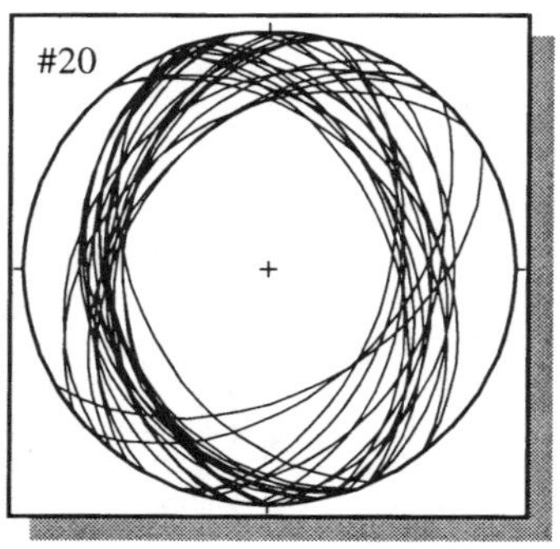

LUNCH SPRING
Straight Cliffs Formation
N = 36

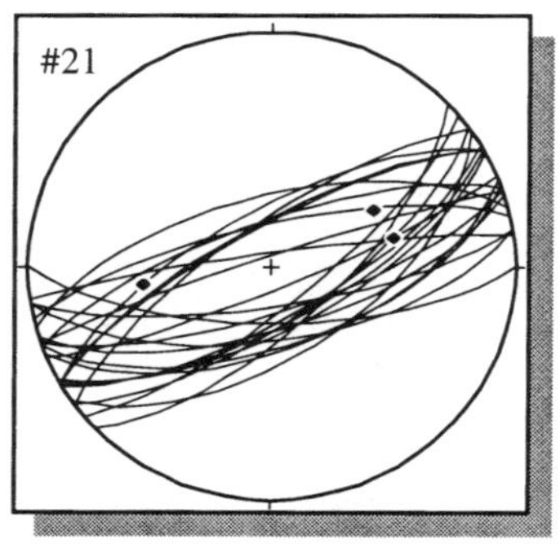

TOP LOCALITY
Upper Navajo Sandstone
N = 24

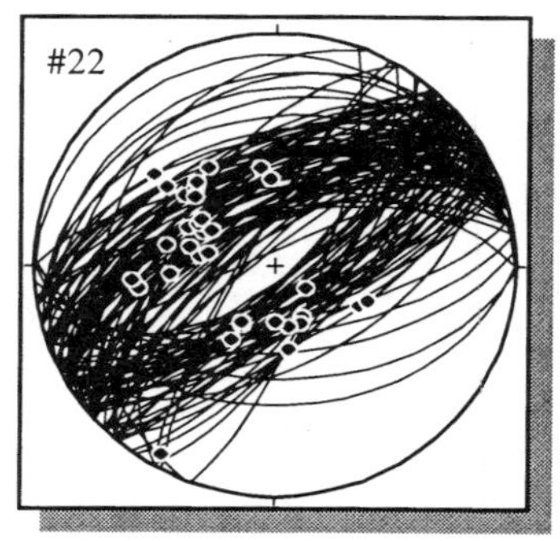

LONG TRAVERSE
Upper Navajo Sandstone
N = 131

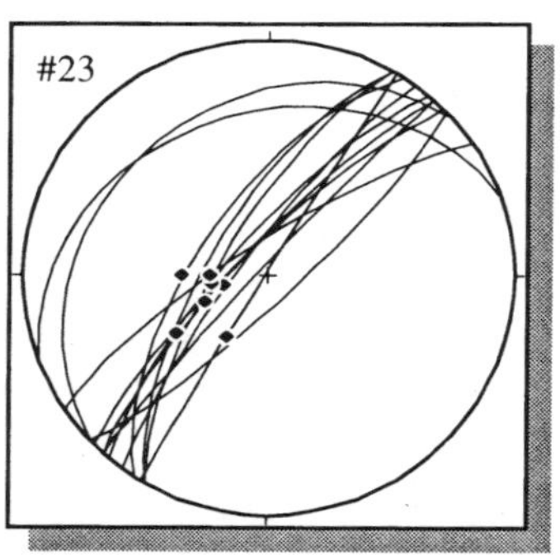

BUTLER VALLEY DRAW
(Reverse Bands)
Navajo Sandstone
N = 13

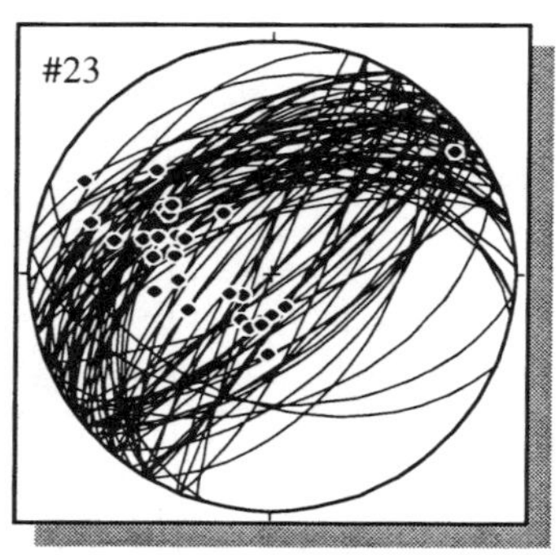

BUTLER VALLEY DRAW
Navajo Sandstone
N = 100

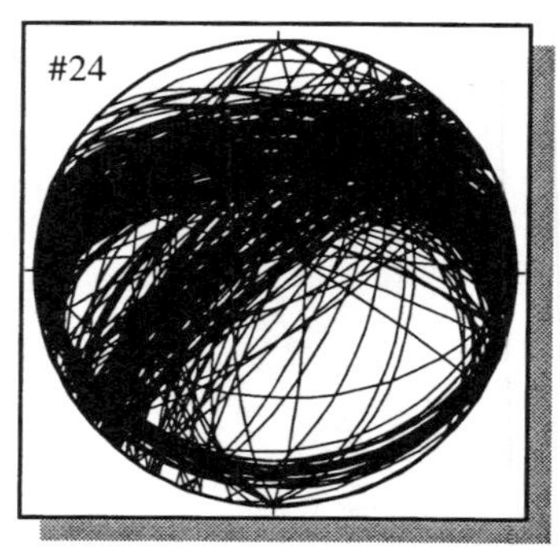

ANGELA'S PLACE
Uppermost Navajo Sandstone
N = 421

B

Figure 85 (this and following page). B: Stereographic projections of deformation bands (great circles) and slickenlines (dots) measured in the Kaibab subregion. Numbers 18–30 refer to the subareas located in A. *N* = number of observations.

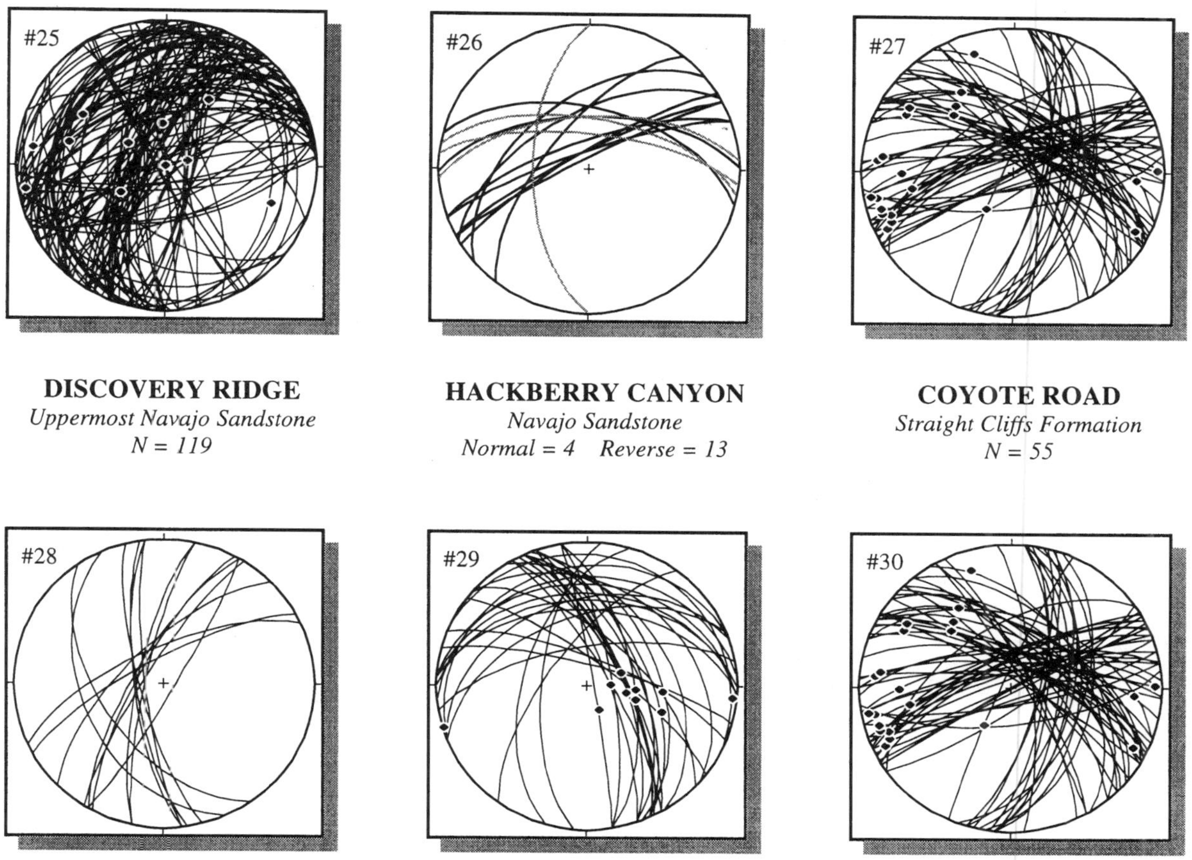

DISCOVERY RIDGE
Uppermost Navajo Sandstone
N = 119

HACKBERRY CANYON
Navajo Sandstone
Normal = 4 Reverse = 13

COYOTE ROAD
Straight Cliffs Formation
N = 55

STRANGELOVE
Straight Cliffs Formation
N = 15

WEST SAND COVE
Uppermost Navajo Sandstone
N = 34

POWER LINE
Navajo Sandstone
N = 69

B

Most of these deformation bands range in strike from ~N45°W to N10°E, dipping either northeast or southwest (Fig. 85B). Some deformation bands in these outcrops, on the other hand, strike east-northeast and dip either northwest or southeast. Dip magnitudes range from ~45° to 80°. Slickenlines are essentially absent, as are slickensided surfaces. Moreover, there are no ladder structures whatsoever.

At the top of the Navajo Sandstone all along the East Kaibab monocline, there are good exposures of the noncataclastic DB2a subvariety of deformation bands (Fig. 87). Almost every outcrop in the uppermost 15 m (50 ft) of the Navajo Sandstone is pervaded by very thin (only as thick as several millimeters), very closely spaced, wispy, irregularly curviplanar deformation bands that occur in multiple conjugate sets and that accommodate nearly imperceptible (millimeter scale) normal displacements. These noncataclastic deformation bands form dense networks. Most outcrops of such deformation bands reveal *multiple* conjugate-normal sets, reasonably systematic in strike directions within a given outcrop, yet with such swings in strikes from outcrop to outcrop that the resultant strike rosettes cover the compass. No ladder-structure geometries are associated with these deformation bands. Slickenlines and slickensides are absent as well. Thin sections reveal that porosity of the host sandstone is commonly diminished within these noncataclastic DB2a deformation bands, but that this loss has been accomplished without cataclasis and with a minimum of grain-scale fracturing. In some outcrops, it is evident that the DB2a deformation bands are overprinted by cataclastic DB1 deformation bands and zones of deformation bands marked by slickenlines, slickensides, and ladder structures.

The Discovery Ridge subarea (25 in Fig. 85A) contains excellent examples of the outcrop characteristics of the DB2a subvariety of deformation bands in Navajo Sandstone. The outcrops reside just beneath the contact between the Navajo Sandstone and the Carmel Formation. Moreover, the Discovery Ridge subarea lies within the middle limb of the East Kaibab monocline, where the bedding dips eastward at ~20°. The bedding surfaces display deformation bands that have anastomosing, inosculating appearances, whereas the vertical sections are marked by relatively straight traces. The deformation bands are very closely spaced (10 to 20 cm [4 to 8 in.]) and display trace lengths that are no more than 2 or 3 m (7 to 10 ft). The outcrops are sharp and hackly, for the

Figure 86. A: South-directed view of the western margin of the Kaiparowits basin. Note the East Kaibab monocline, which is represented by east-dipping Straight Cliffs Formation and Wahweap Sandstone (Upper Cretaceous). B: North-directed view of the edge of the Kaiparowits basin as well as the east-dipping Straight Cliffs Formation (in foreground) along the East Kaibab monocline. Note deformation bands in foreground.

deformation bands project out of the outcrops as sharp, resistant, thin blades and points, with relief of several to 10s of centimeters. Conjugate patterns are the rule, and they can be seen on cross-sectional outcrop faces of almost any orientation (Fig. 87). Stereographic projections of measurements of such deformation bands tend to produce "ball of string" diagrams, such as that representing the orientations of deformation bands in the Discovery Ridge subarea (25 in Fig. 85B). In places within this subarea, it is possible to clearly recognize the presence of cataclastic deformation bands and zones of deformation bands, complete with ladder structures, slickenlines, and slickensides, superimposed upon the DB2a subvariety of noncataclastic deformation bands. Also evident in outcrop and in stereographic projection (25 in Fig. 85B) is the predominant west dip direction of the deformation bands, ranging from shallow to steep. These dips reflect the rotation of bedding (during monoclinal folding) of the DB2a deformation bands, which originally occupied *horizontal* Navajo Sandstone. Even in the now-tilted mode, the acute bisector of two conjugate DB2a deformation bands is always approx-

Figure 87. North-directed photograph of the very top of the Navajo Sandstone along the East Kaibab monocline. Bedding dips ~15°E. DB2a deformation bands (traced in white) are conjugate normal, and their acute bisectors are perpendicular to bedding. These deformation bands formed when the Navajo Sandstone was essentially horizontal. Subsequently they were rotated, along with bedding, during monoclinal folding.

Figure 88. A: Large hand sample of deformation band in Navajo Sandstone. The upper surface of this sample is polished (slickensided) smooth as glass, and marked by slickenlines. Cross-sectional view shows contrast between deformation band (white, frosted) and host rock. B: Close up view of deformation band (white) and normal texture of Navajo Sandstone.

imately perpendicular to bedding (Fig. 87). The millimeter-scale offsets sometimes observable are consistent with the interpretation of compaction during burial and stratigraphic loading. Because bedding is inclined at ~20°E in this subarea, some deformation bands that were originally steeply east dipping and marked by normal offset are now rotated into west-dipping orientations marked by reverse offset. Those that originally dipped steeply westward now dip more shallowly westward.

Properties of the superposition of cataclastic and noncataclastic deformation bands

The contrast between the DB2a subvariety of deformation bands in uppermost Navajo Sandstone with the DB1 variety of deformation-band shear zones of tectonic origin is unmistakable. The outer margins of deformation-band shear zones of tectonic origin commonly are marked by slickenlines and slickensides (Fig. 88). In overstepping regions, ladder structures are well developed. Microfabrics are cataclastic in nature, marked by the combination of grain fracturing, cataclastic flow, and collapse of porosity.

A good example of the superposition of DB2 deformation bands by DB1 deformation bands is seen at Angela's Place subarea (24 in Fig. 85A). Angela's Place subarea lies stratigraphically near the very top of the Navajo Sandstone, and the surrounding landscape is marked by the presence of "fault fins," expressions of the resistance of the tectonic deformation-band shear zones to weathering and erosion (Davis, 1998). Some of the fins in this part of the East Kaibab monocline are as tall as 8 m (26 ft) (Fig. 89). Angela's Place subarea is a single, thick, deformation-band shear zone that has been weathered and winnowed to such an extent that the internal architecture of the deformation-banded rock is clearly revealed (Fig. 90). Porous sandstone occupying locations between the closely spaced deformation bands and zones of deformation bands is winnowed out, leaving a honeycomb appearance that we refer to as "radiator rock," a coin termed by Randy Tufts (personal commun., 1995). The complexity of the patterns reflects a superposition of cataclastic, tectonic DB1 deformation bands on noncataclastic, pretectonic DB2a deformation bands. The DB2a deformation bands are rotated from their original orientations by bedding tilt accompanying monoclinal folding. The key to interpreting the pattern of deformation at Angela's Place was found by tracing thick deformation-band shear zones downward into stratigraphically deeper parts of the Navajo Sandstone and observing the change in physical characteristics. Beneath the upper 15 m (50 ft) of the Navajo Sandstone, the zones of deformation bands

Figure 89. Large (8 m [26 ft] tall) fault fin, whose geomorphic expression is controlled by the resistance of a deformation-band shear zone of tectonic origin. The deformation-band shear zone protects and thus reduces the rate of weathering and erosion of footwall cross-bedded sandstone. Geologist is Danielle Vanderhorst.

Figure 90. A: North-directed view of weathered deformation-band shear zone at Angela's Place subarea. Angela Smith is the geologist. Note eastward dip of beds in background. B: West-directed view of the same outcrop area. This view captures the expression of moderately steeply north-dipping deformation bands, which strike at a high angle to the zone as a whole. Angela is still at work at the top of the outcrop.

have a much simpler architecture, in the form of DB1c ladder-structure geometries, unencumbered by the presence of the earlier record of DB2a deformation bands.

To be sure, the presence of the DB2a deformation bands has a significant effect on the outcrop appearance of the superimposed cataclastic deformation-band shear zones, especially where they crop out as "fault fins" because of their resistance to erosion. In particular, where thick deformation-band shear zones of a cataclastic origin are seen at the stratigraphic level of uppermost Navajo Sandstone, the outcrops are pervaded by very closely spaced, limonite- and manganese-stained, resistant deformation bands. The outcrops contain the combination of conjugate-normal, noncataclastic deformation bands and synthetic and antithetic deformation bands of a cataclastic nature that form ladder structures in the superimposed tectonic deformation-band shear zone. The outcrops have a very complex aspect especially where the sandstone has been winnowed into "radiator rock." They take on a scabby appearance.

Properties of the cataclastic deformation-band shear zones

The upper hinge region of the East Kaibab monocline, and the upper part of the middle limb, are marked by an extremely well developed system of deformation-band shear zones. On the upland surfaces between canyons, the zones of deformation bands express themselves both as long, resistant, continuous dike-like structures and as short, discontinuous blades or fins, quite dramatic in form (Fig. 91). The overall physiographic expression is a "fault-fin landscape" (Davis, 1998), shaped on the upland surfaces where there has been plenty of time for weathering and erosion to reveal the dramatic contrasts between the resistance of porous sandstone host rock and resistance of the deformation-band shear

Figure 91. Photograph of especially well developed, resistant, deformation-band shear zones. Tallest blade is ~12 m (39 ft).

Figure 92. Photograph of examples of resistance of deformation-band shear zones to erosion. We call these the "flying pancakes." The column of sandstone beneath the pancake on the right is ~2.5 m (8 ft) tall and represents the footwall of the deformation-band shear zone. The footwall sandstone has remained porous and is of course much less resistant to erosion than is the deformation-band shear zone itself.

Figure 93. Photograph of profile view of a deformation-band shear zone, which protects underlying footwall sandstone from erosion. Note the triangular form. Day pack for scale.

zones (Fig. 92). The landscape is typically marked by short (1 m [3.3 ft]) to tall (10 m [33 ft]) fins and blades that project upward from the land surface, or much more rarely, outward from canyon walls. In cross section, the fins are commonly triangular; their apical angles are ~50°, and one of the sides of the triangle project above and beyond the other (Fig. 93). The flanks of the fins, which together define the triangular cross-sectional forms, are asymmetrical in their physical make up, one side composed of the sandstone host, complete with primary sedimentary structure such as bedding, cross-bedding, and primary deformational features. The other side, which may project higher, is devoid, or nearly so, of any expression of the primary stratification of the sandstone.

Figure 94. Longitudinal profile view of deformation-band shear zone. Note the scalloped appearance of upper surface. Day pack for scale.

Figure 95. Photograph of L-structure (the structure grain raking from upper left to lower right) on flank of deformation-band shear zone. Geologist is Danielle Vanderhorst.

Figure 96. Southwest-directed "virtual" oblique view of the Cottonwood Creek area along the northern part of the East Kaibab monocline. Note dike-like expression of deformation-band shear zones. This image was created by Chevron, Inc., through draping our large-scale aerial photographic imagery over our digital topography.

Viewed longitudinally, the tops of the fins have a scalloped appearance, and not uncommonly the fins are marked by windows (crudely circular) or breached windows (Fig. 94). The outer surfaces of deformation-band shear zones commonly display slickenlines, slickensides, and L-structure (Fig. 95). Furthermore, the surfaces are commonly stained by manganese and marked by lichen growth. In the canyon bottoms, the deformation-band shear zones also project upward or outward from bedrock because of resistance to erosion, except in the bedrock drainage channels proper, where they are typically eroded flush with the overall bedrock. However, the thick deformation-band shear zones along the canyon bottoms clearly control the locations of dry waterfalls and plunge pools. Moreover, at several different locations, the confluence of two tributary drainages coincides with the location of a resistant zone of deformation bands oriented at a high angle to the flow of the main drainage, as if the zone created an impediment to headward erosion by the main branch, causing it to split into two tributary drainages.

Not surprisingly, the expression of these walls and fins of deformation-band shear zones show up beautifully in aerial photographs. Figure 96, for example, is a southwest-directed, oblique *virtual* image of the fault-fin landscape near the northern end of the Kaibab uplift. The image was created by colleagues at Chevron (see Acknowledgments) who draped my GPS-controlled 1:600 topographic base map (10 ft [3.3 m] contour intervals) with the 1:600 scale aerial photographs flown especially for this study, which were "rubber-sheeted" to correct somewhat for distortion. The virtual image captures the dike-swam–like nature of the deformation-band shear zones, which strike northeast and dip steeply. Many are hundreds of meters in trace length and several meters thick. The largest such deformation-band shear zone, which is up to 6 m (or 20 ft) thick, can be traced for more than 1 km (0.6 mi).

The tectonic nature of these deformation bands, which create the fault-fin landscape, is quite obvious in outcrop because of the presence of slickenlines and slickensides. Moreover, microscopic examination of thin sections of the deformation-band shear zones reveals grain-size reduction achieved through grain fracturing and cataclastic flow (Fig. 97). The deformation-band shear zones of cataclastic origin have accommodated offset, ranging from fractions of centimeters to tens of meters. Cross-bedding in the Navajo Sandstone provides ubiquitous markers for assessing offsets, although from place to place the determination of offset is challenging indeed! Unpublished data generated in this study and describing deformation-band shear-zone thickness vs. amount of offset suggest that offset generally is 2× to 8× the zone thickness.

Mapping of the cataclastic deformation-band shear zones

My students and I (see Acknowledgments) mapped the system of these deformation-band shear zones along the northern part of the East Kaibab monocline at two different scales, first by using 1:12,000 color aerial photographs and second by using 1:600 black and white aerial photographs flown especially for

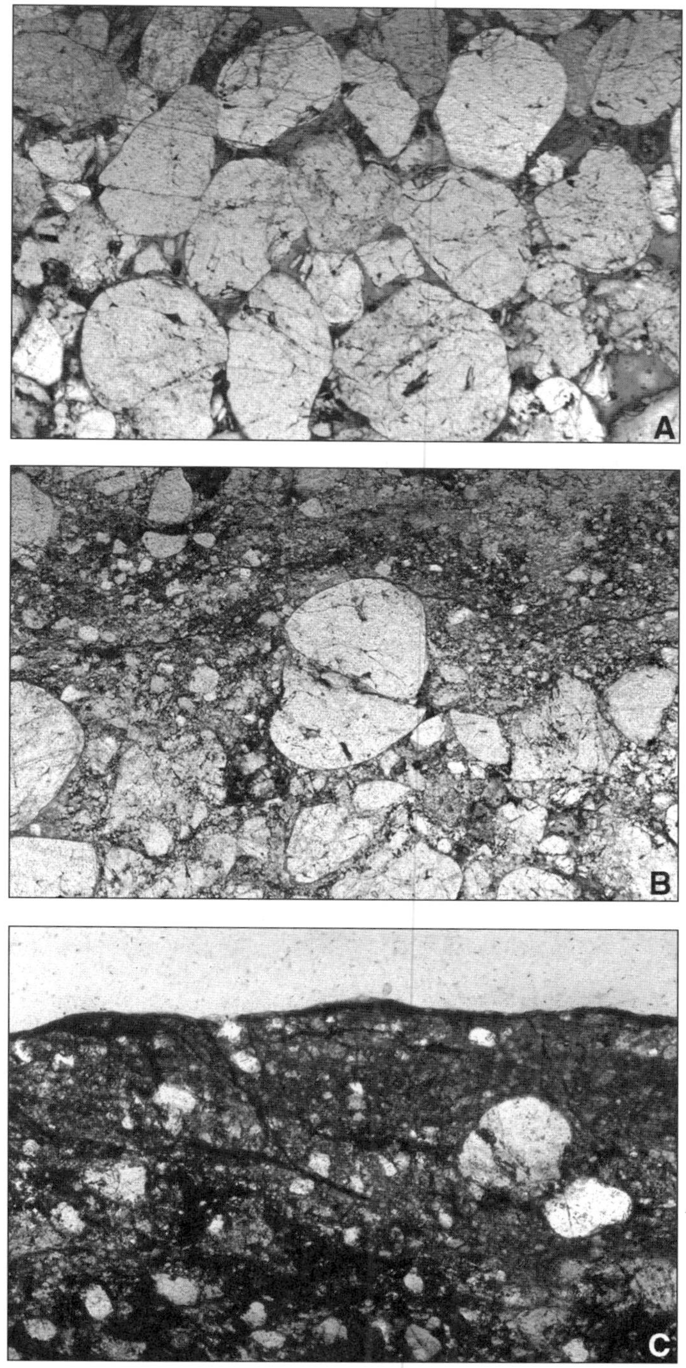

Figure 97. Photomicrographs of Navajo Sandstone, revealing the nature of progressive microfracturing and cataclasis during the formation of deformation bands. Specimens collected from the Kaibab and Waterpocket subregions. A: Undeformed grains in a porous framework. Width of field of view is 320 μm. B: Deformation band marked by very fine grained comminuted quartz. Note that a microfault has cut and displaced one of the surviving sand grains. Width of field of view is 320 μm. C: The extremely fine grained comminuted rock along the top margin of this thin section corresponds in the field to a slickensided "rind" on the outermost margin of a deformation-band shear zone. The coarser-grained comminuted rock is the deformation band itself. Width of field of view is 320 μm.

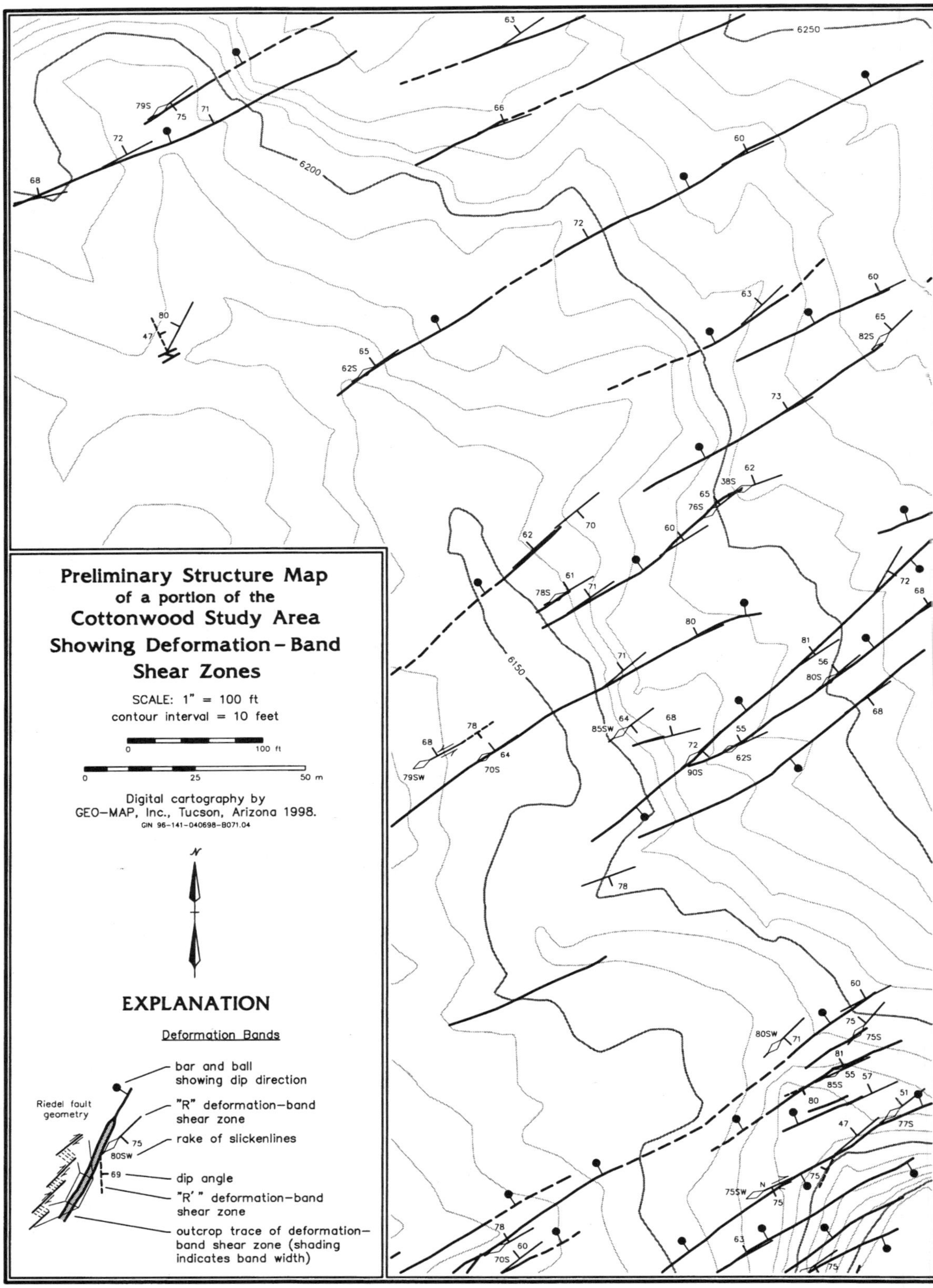

Figure 98 (this and next page). A sample of the structural geologic mapping carried out in the Cottonwood area along the East Kaibab monocline.

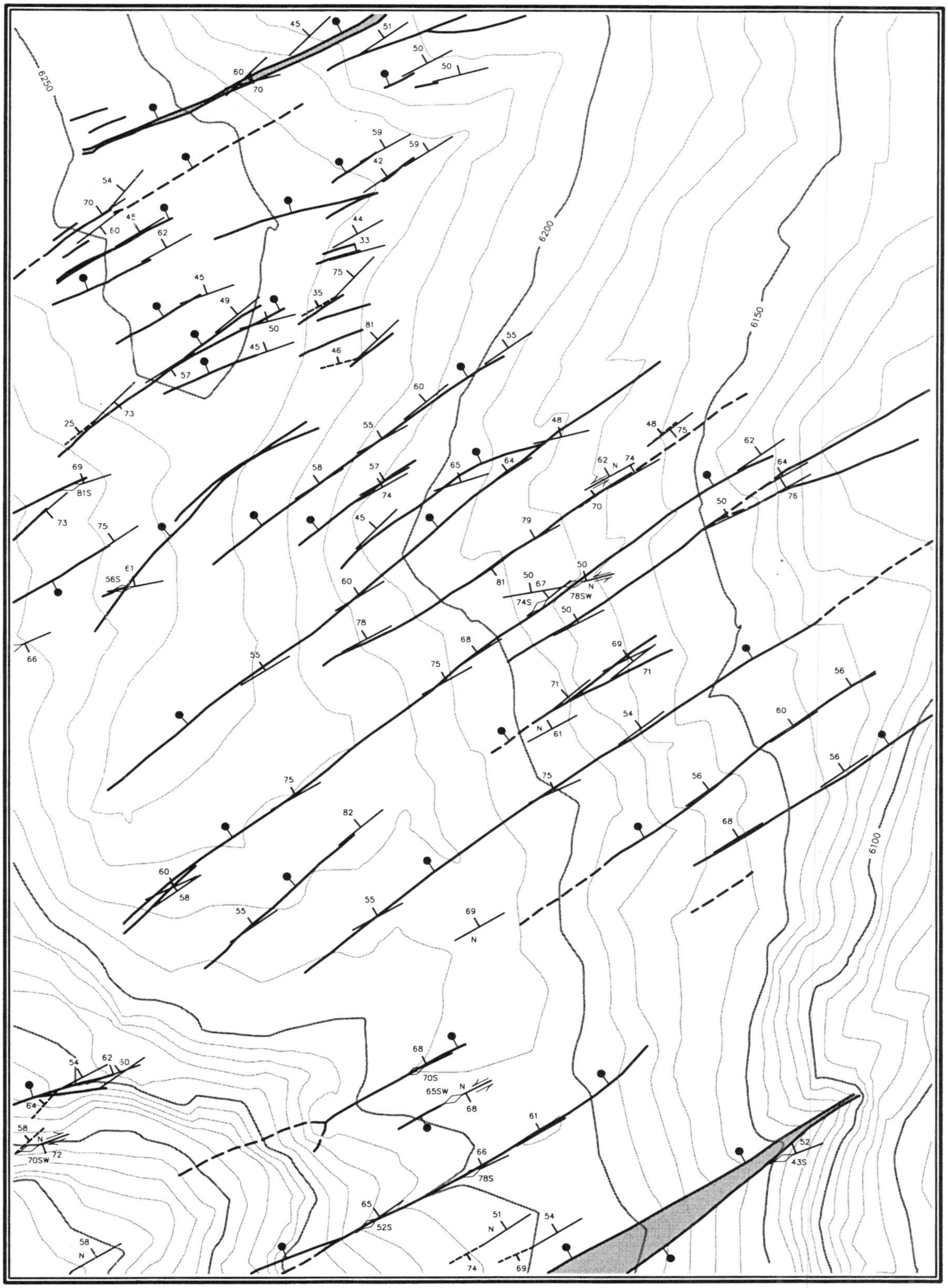
6250
6200
6150
6100

this project. Outcrops are outstanding. The steep-sided canyon exposures permit the critical third dimensions of the deformation-band shear zones to be observed and mapped. Accurate placement of contacts was afforded by the high resolution of the aerial photographs. Because of the shadows cast by even the smallest projecting features, we were able to resolve deformation-band shear zones as thin as 10 cm (4 in.). As a consequence, we were able to map essentially *all* of the deformation-band shear zones exposed in the area.

A sample of the results of the mapping is presented in Figure 98. The deformation-band shear zones strike northeast and dip moderately to steeply both to the northeast and to the southwest. Distance between thick (1 to 5 m [3.3 to 6.5 ft]) deformation-band shear zones commonly ranges from 10 to 20 m (33 to 66 ft). The deformation-band shear zones disclose normal separation, except for those that lie along the easternmost margin of the map area, where the dip of bedding of the host Navajo Sandstone increases to greater than ~10°E. There, some show reverse offset. Slickenlines and L-structure on outer surfaces of the deformation-band shear zones consistently rake steeply southwestward, revealing that a small component of the displacement is strike slip in nature.

During the course of the mapping we developed a much better understanding of the connectivity of the deformation-band shear zones (García and Davis, 1997). Each of the mapped deformation-band shear zones shown in Figure 98 consists of relatively thin (3 to 50 cm [1 to 19 in.]), massive, continuous bands (Fig. 99) that are connected by ladder-structure transfer zones where the bands overstep one another (Fig. 100). The massive, continuous bands resemble quartzite layers of either uniform thickness or slowly, uniformly changing thickness (Fig. 99). Thus each mapped trace of a long, continuous deformation-band shear zone consists of a linked array of multiple, continuous, massive deformation-shear zones and multiple transfer zones. Almost every exposure of transfer zones shows well-defined hanging-wall and footwall zones and surfaces, not uncommonly marked

Figure 99. A: Photograph of long, continuous, massive variety of deformation-band shear zone in the Cottonwood area. Scott Witthoft is the geologist. B: Photograph of an even larger example. Notice the small antithetic band branching from the wall. Geologists are Pilar García and Scott Witthoft.

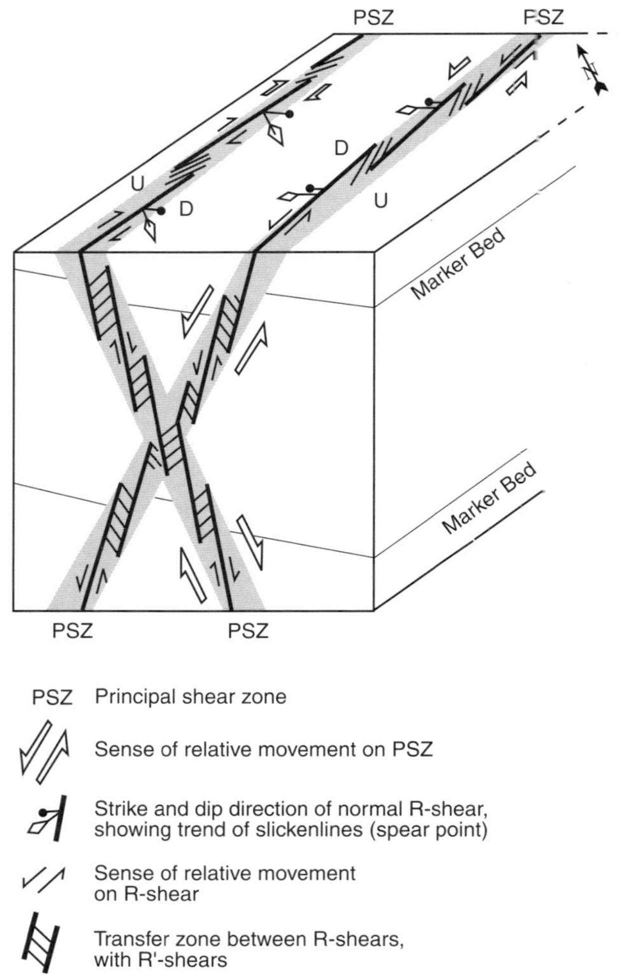

Figure 100. Line drawing showing long, continuous, massive bands that overstep at transfer zones. Note that the spacing between transfer zones in this normal-fault system is much closer in cross-sectional view than in plan view.

by patches of slickensides and slickenlines, between which lay a winnowed, weathered boxwork zone composed of ribs of deformation bands. The deformation bands that are parallel to the hanging-wall and footwall surfaces of the transfer zones are distinguished by motions and offsets that are synthetic with those of the zone as a whole (Figs. 98 and 100). The others, which typically make angles of ~50° to 80° to the hanging-wall and footwall surfaces, are distinguished by motions and offsets that are antithetic to those of the zone as a whole (Figs. 98 and 100). Slickenlines are oriented perpendicular to the intersection of the synthetic surfaces and the antithetic surfaces. Sense of offset for the whole zone is disclosed by the polarity of the acute angle between synthetic and antithetic deformation bands (Fig. 100).

Until these relationships were worked out, it was difficult to confidently assess the sense of offset on deformation bands in the map area, in spite of the fact that the Navajo Sandstone host rock contains abundant cross-bedding that can be used as markers. Part of the difficulty was simply related to the character of exposure on the upland surfaces, where the deformation-band shear zones, without much hanging-wall and footwall host rock, project up from the subhorizontal bedrock surface and thin soil cover. Another part of the difficulty, however, was related to kinematic contradiction: Almost all of the deformation-band shear zones in the area show essentially normal slip, yet some in the easternmost part of the mapped area show reverse slip. This apparently contradictory observational information sorted itself out when it became obvious that *all* are normal and extensional with respect to overall bedding orientation. Only where bedding rolls over into moderate eastward dips within the monocline proper do high-angle reverse-offset and lower angle normal deformation bands appear (Fig. 101).

It appears that the fundamental origin of these cataclastic (DB1) deformation-band shear zones was related in part to extensional stretching on the outer arc of the upper hinge of the monocline, in such a way that there likely is a close relationship between degree of faulting and amount of curvature. This interpretation is supported by the fact that the deformation-band shear zones are only found in the upper-hinge region of the East Kaibab monocline. The deformation bands must have begun forming at the very onset of bending, because when the East Kaibab monocline is unfolded, the pattern of this deformation-band shearing is one of conjugate-normal faulting. The obvious difficulty with this overall interpretation is the obliquity between the ~N20°E trend of the East Kaibab monocline and the ~N40°E strike of the deformation-band shear zones. Tindall and Davis (1999) interpreted this difference to be related to the transpressive, reverse, right-lateral origin of the East Kaibab monocline (Fig. 102).

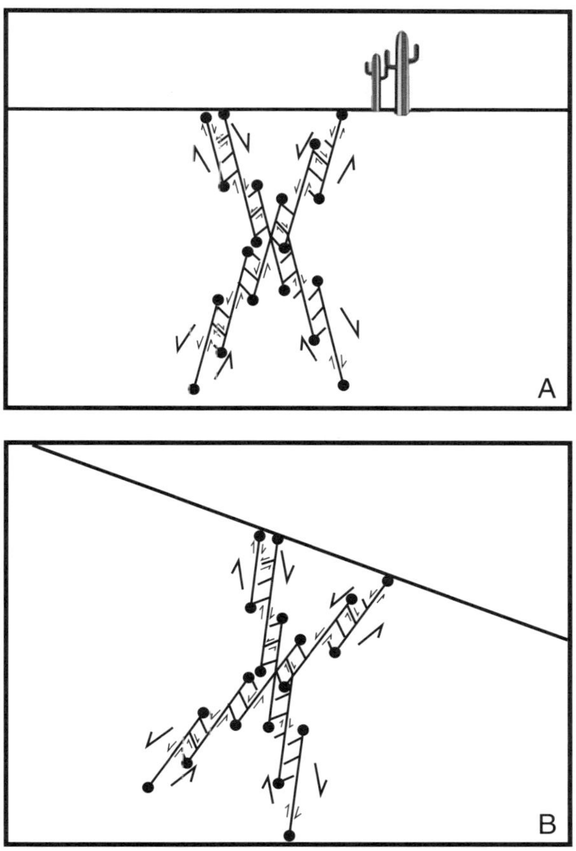

Figure 101. Conjugate-normal deformation-band shear zones reveal layer-parallel extension, no matter whether bedding is (A) horizontal or (B) inclined.

Recognition of Riedel shearing

The individual mapped traces of the massive and continuous deformation-band shear zones do not coincide exactly with the longer map trace of the deformation-band shear zones of which they are a part (Figs. 98 and 100). The detailed mapping makes it obvious that the resistant zones of deformation bands that appear to be so "continuous" at one scale actually are discontinuous Riedel (R) segments linked by R′ segments in transfer zones (Davis, 1996b; Davis et al., 1999). The transfer zones occupy areas where the R segments overstep one another (Fig. 101). Consistent with classic Riedel geometries, the R segments make an angle of ~15° degrees with the zone of which they are a part. This

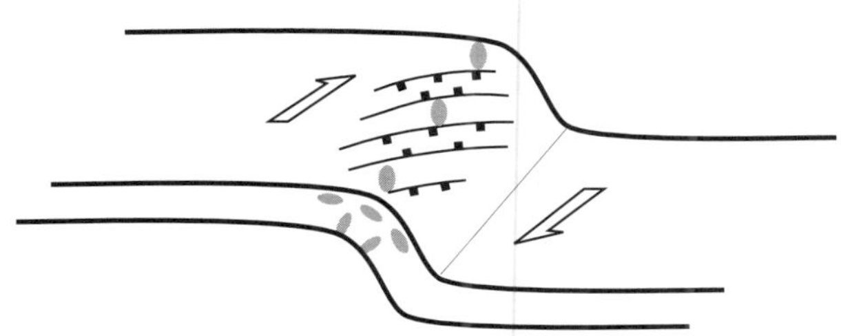

Figure 102. Line drawing suggesting that the orientations of the deformation-band shear zones along the northern part of the East Kaibab monocline are consistent with folding during transpressive, reverse, right-lateral shearing. The ellipses are strain ellipses.

Figure 103. A: Photograph of canyon-wall outcrop of deformation-band shear zones. B: The approximate trace of the map expression of the deformation-band shear zone is shown as a wide, very light gray streak. The individual Riedel shears are marked with lines and arrows.

pattern is demonstrated by (1) carefully examining the strike-and-dip attitudes of deformation-band shear zones from place to place along a given resistant zone of deformation bands and (2) noting the discordance and the systematics of discordance. Even more revealing, it is useful to carefully study canyon-wall exposures of these zones of deformation bands, where the Riedel patterns for *normal* faulting are most obvious and where it can be shown that the width of the resistant, continuous zone itself is defined by the tip lines of the stepped Riedel shears. This pattern is pictured schematically in Figure 101. Where the bedding is essentially horizontal, the fundamental Riedel geometric and kinematic relationships are as pictured in Figure 101A. West-northwest–dipping, predominantly normal-slip zones are composed of yet more steeply west-northwest–dipping Riedel shears, and east-southeast–dipping, predominantly normal-slip zones are composed of yet more steeply east-southeast–dipping Riedel shears. In each case, the Riedel shears are overstepped and connected by relays within which R′ shears are abundantly developed. Where bedding dips easterly (Fig. 101B), the west-northwest–dipping, predominantly normal-slip zones, along with the Riedel shears, display lower west-northwest dip angles than where bedding is horizontal. The other set dips steeply west-northwest, with the Riedel shears being shallower by ~15°. The most kinematically provocative arrangement occurs where the dip of bedding is such that steep east-southeast dipping (thus predominantly normal slip) zones contain individual R segments that are west-southwest dipping (thus predominantly high-angle reverse in nature). The evidence for this Riedel geometry is seen where there is excellent three-dimensional exposure, e.g., on the walls of canyons (Figs. 103 and 104).

Further descriptions of properties of deformation bands in Navajo Sandstone in the northern part of the Kaibab subregion

With this background perspective in mind, it is useful to examine the data collected from north to south through the

Figure 104. North-directed view of deformation-band zone in Navajo Sandstone in the Cottonwood study area. Note that individual deformation-band shear zones dip more steeply than the zone as a whole.

Kaibab subregion along the East Kaibab monocline (Fig. 85). Just east of the crest of Butler Valley anticline, at the Top Locality subarea (21 in Fig. 85A), there are well-exposed, cataclastic (DB1) deformation-band shear zones that can be traced for distances greater than 100 m (330 ft). Individual deformation-band shear zones are as much as 2 m (7 ft) thick. The deformation-band shear zones range in strike from ~N50°E to N85°E and dip at ~60° to 70°, both to the northeast and to the southwest. The map relationships reveal that the deformation bands compose a conjugate-normal system. Individual outcrops show DB1a and DB1c subvarieties (see Appendix Table 1) marked by parallel discrete zones and ladder structure, respectively. The deformation bands occur in the upper part of the Navajo Sandstone, where it dips just several degrees southeastward. To the east of Top Locality, the Long Traverse subarea (22 in Fig. 85A) shows nearly the identical pattern, although the strike directions for these conjugate-normal deformation-band shear zones for the most part range from ~N45°E to N75°E. Slickenlines are abundant. They rake steeply southwest no matter whether the dip direction of the deformation-band shear zone with which they are associated dips northwest or southeast. The northwest-dipping deformation-band shear zones show predominantly normal slip but with a slight left-handed strike-slip component, and the southeast-dipping deformation-band shear zones show predominantly normal slip but with a slight right-handed strike-slip component. The dip of Navajo Sandstone in the Long Traverse subarea is, for the most part, less than ~5°SE.

Even farther to the east, within the Butler Valley Draw subarea (23 in Fig. 85A), there are abundant conspicuous fault-fin expressions of DB1a and DB1c subvarieties deformation-band shear zones in Navajo Sandstone, which dip southeastward from ~10° to 12°. Offset is normal, with magnitudes ranging from centimeters to several meters. Only one reverse-offset deformation-band shear zone was noted, and this was found where bedding was most steeply dipping. As shown stereographically (Fig. 85B), the pattern of deformation-band shearing in the Butler Draw (subarea 23) is one of conjugate-normal faulting, similar to that of the Top Locality subarea (21 in Fig. 85A) and the Long Traverse subarea (22 in Fig. 85A), but again there is a rotation toward the north of the average strike attitudes. In the Butler Valley Draw subarea, the strike of the conjugate-normal deformation-band shear zone ranges from ~N30°E to N70°E. The relationship of bed tilt (~10°) to the dip of the conjugate-normal deformation-band shear zones is clear stereographically, for the southeast-dipping set of deformation-band shear zones in the Butler Valley Draw subarea (23 in Fig. 85B) is much steeper than the northwest-dipping set. Where the Navajo Sandstone bedding is steepest, a few of the originally normal-offset deformation-band shear zones are reverse in nature. As in the Long Traverse subarea (22 in Fig. 85B), slickenlines rake steeply southwest. Thus of the two sets of conjugate deformation-band shear zones in the Butler Valley Draw area, the southeast-dipping set is high-angle normal with a slight right-handed component, and the northwest-dipping set is moderate- to low-angle normal with a slight left-handed component. Many of the data shown for Butler Valley Draw subarea (23 in Fig. 85B) were collected about 75 m (247 ft) beneath the contact between the Navajo Sandstone and the Carmel Formation.

At Angela's Place subarea (24 in Fig. 85A), where the Navajo Sandstone dips ~20°SE, the stereographic pattern of measurements can be better appreciated. The predominant outcrop character of deformation bands at Angela's Place (Fig. 90) is ladder structure (DB1c) within a reverse-offset deformation-band shear zone that is 2 m (7 ft) thick. It strikes ~N50°E and dips ~80°W. The bold swath of great circles striking ~N40° to 60°E and dipping steeply west corresponds to the synthetic set of deformation bands, whose kinematics are reverse with a very small right-handed component. The density of these data reflects the high number of synthetic deformation-band shear zones within the outcrop. The bold swath of great circles striking approximately east and dipping gently to moderately northward (24 in Fig. 85B) is produced by measurements of the numerous antithetic deformation bands within the outcrop. The background "noise" within the stereographic projection is created primarily by the noncataclastic (DB2a) conjugate-normal deformation bands of the type that proliferate in the uppermost 10 to 15 m (33 to 50 ft) of the Navajo Sandstone. When the stereographic pattern is rotated in ways that conform to rotating bedding to horizontal, the synthetic and antithetic elements restore to a straightforward normal-offset arrangement. A similar pattern may have been sampled, but not nearly as densely, in the Hackberry Canyon subarea (26 in Fig. 85A). Bedding in the Navajo Sandstone dips ~30°SE in the Hackberry Canyon subarea, and the distribution of attitudes of normal vs. reverse deformation-band shear zones is like that of Angela's Place.

The Discovery Ridge subarea (25 in Fig. 85A) is one of especially well-developed noncataclastic (DB2a) deformation bands in Navajo Sandstone. The scatter of orientations that typifies this class of deformation bands is apparent stereographically (25 in Fig. 85B) and is strikingly different from the stereographic patterns of the DB1a and DB1c subvarieties of cataclastic deformation-band shear zones that occur in subareas 21, 22, and 23 (Fig. 85B). Dips of the DB2a deformation bands in the Discovery Ridge subarea are almost exclusively westward because Navajo bedding is dipping ~20°E.

At Candyland, to the east of the Discovery Ridge subarea, lies the slot canyon that my students and I have called "Butterfly Canyon." The canyon exposes steep limb dips (as steep as ~65°E) of the East Kaibab monocline. The deformation-band structures in these more steeply dipping parts of the East Kaibab monocline are very difficult to characterize, for the patterns reveal complex products of progressive deformation. The DB1d subvariety of deformation bands is abundant in the slot canyon, forming parallel to bedding and cross-bedding. These appear to reflect a flexural-slip reactivation of bedding during monoclinal folding. The physical expression of such deformation includes resistant bedding-parallel bands or zones of bands that are especially white and vein-like and that tend to project outward from the outcrop face. It is logical that these would occur, given the magnitude of bedding-parallel slip that is demanded by the folding. In addition there are obvious thrust-slip cataclastic (DB1a) deformation-band shear zones, and these tend to form in the lower parts of the monoclinal fold just above the lower hinge. Some of the these thrusts are associated with overturned folds whose axial-plane regions may in turn may be marked by zones of kink folds. However, there are many deformation bands in Butterfly Canyon that do not conform to simple generalizations, for even a quick walk through the slot canyon will reveal normal-, thrust-, reverse-, and strike-slip offset. It will be necessary to compare the nature of the kinematics and dip direction(s) of individual deformation-band shear zones to the overall dip magnitude of bedding in order to try to access the critical matter of the orientation of the structure at the *initial* time of its development. Boardman (1995) has used this approach with some success in studying deformation bands in this general area.

Properties of deformation bands in Straight Cliffs Sandstone in the northern part of the Kaibab subregion

It is informative to compare the record of deformation-band shear zones in the Navajo Sandstone (Jurassic) with the record in the Straight Cliffs Formation (Upper Cretaceous) in essentially the same structural position on the East Kaibab monocline. In the northern part of the Kaibab subregion, the exposures of Straight Cliffs Formation are east and up-section from those of the Navajo Sandstone. The Gut Road, which branches from the Cockscomb Road, provides access to an extraordinary panorama of the Kaiparowits basin as well as a view of its western boundary in the form of east-dipping Straight Cliffs Formation along the East Kaibab monocline (Fig. 86). Although most of the lithologies within the Straight Cliffs Formation and overlying Wahweap Sandstone do not accommodate strain through the formation of deformation bands and zones of deformation bands, there is one silvery white unit of porous sandstone that is clearly an exception. It is replete with deformation bands and zones of deformation, with density and apparent complexity of deformation banding that are almost overwhelming (Fig. 86B). Fortunately, these deformation-band shear zones are cataclastic (DB1) and of tectonic origin and thus are commonly marked by slickenlined, slickensided surfaces (Fig. 105) as well as ladder structures. The geometries serve as guides to evaluating sense of shear. To my knowledge, there are no primary-compaction deformation bands of noncataclastic origin within this array of structures.

The Cad's Crotch subarea (19 in Fig. 85A) is an easily accessible and productive area for analyzing this system of deformation bands and zones of deformation bands. This subarea is located close to major reverse right-handed faults that offset units within the Upper Cretaceous by distances of 0.5 to 1 km (0.3 mi to 0.6 mi) (19 in Fig. 85A; also see Fig. 36). Bed-

Figure 105. Photograph of resistant deformation-band shear zone marked by slickenlines and slickensides. Note dark limonite staining on parts of the surface of the zone. Host rock is Straight Cliffs Formation (Upper Cretaceous). Geologist is Sarah Tindall.

ding in this subarea typically strikes ~N20°E and dips from ~25° to 40°ESE. Figure 106 schematically reveals my interpretation of the pattern of deformation-band shear zones in the Cad's Crotch subarea. The stereographic presentation of the pattern (19 in Fig. 85B) resembles a "ball of string." Within the pattern there are steeply northeast-dipping, northwest-striking, predominantly left-handed deformation bands and steeply north-dipping, east-northeast–striking, predominantly right-handed deformation bands. In addition there are abundant north-striking, east-dipping deformation bands. My impression is that there are two conjugate sets of oblique-slip deformation-band shear zones, and each is marked by ladder structure. Together they accommodated a shortening approximately perpendicular to the north-northeast trend of the East Kaibab monocline (Fig. 106A). An additional complexity emerges from out-of-the syncline, westward thrust slip, up the dip of the Straight Cliffs Formation (Fig. 106B). The movement is a response to a space problem in the tightly flexed syncline just east of the East Kaibab monocline. This movement created a third shear set, best viewed in cross section (Fig. 106B). Significantly, the synthetic and antithetic elements of these bedding-parallel DB1d zones are essentially parallel to the strike of bedding, signifying that movement is directly up the dip direction of bedding.

Oblique slip is thus the characteristic of deformation-band kinematics in the Cad's Crotch subarea. This finding perhaps is not surprising given the fact, as discussed by Tindall and Davis (1999), that a major fault of the East Kaibab monocline breaks northeastward from its ordinary monocline-parallel N20°E path, cutting up-section along splays that strike ~N45°E and dip 75°NW. Major reverse right-handed faults offset Upper Cretaceous units by distances of 0.5 to 1 km (0.3 to 0.6 mi). Bedding at this locality typically strikes ~N20°E and dips from 25° to 40°SE. Two of these fault splays pass in close proximity to the Cad's Crotch subarea. Close inspection of these faults reveals the slickenlines and grooves that rake ~15° to the southwest, thus recording reverse right-handed movement. As is clear from map patterns, the entire outcrop trace of the Straight Cliffs Formation bends in a broad, sweeping right-lateral way just to the south of the break-away zone of the right-handed master fault (Figs. 36 and 85A).

Another example of this same behavior of the Straight Cliffs Formation is found at the Coyote Road subarea (27 in Fig. 85A), in the *exact* structural position as that of the Cad's Crotch subarea but some 30 km (18 mi) to the south. Deformation bands abound! Bedding in the Straight Cliffs Formation strikes ~N25°E and dips ~40°SE. The zones of deformation bands express themselves as cataclastic ladder structures (DB1c), and some layer-parallel beds appear to represent thrust-slip shear up the dip of the sandstone (i.e., westward out of the syncline). Stereographic data for the Coyote Road subarea (27 in Fig. 85B) show steeply east-dipping reverse-slip deformation bands that strike north-south; right-handed shear zones strike east-northeast, and left-handed shear zones strike northwest. The shortening implied by these conjugate deformation-band shear zones is approximately perpendicular to the trend of the East Kaibab monocline.

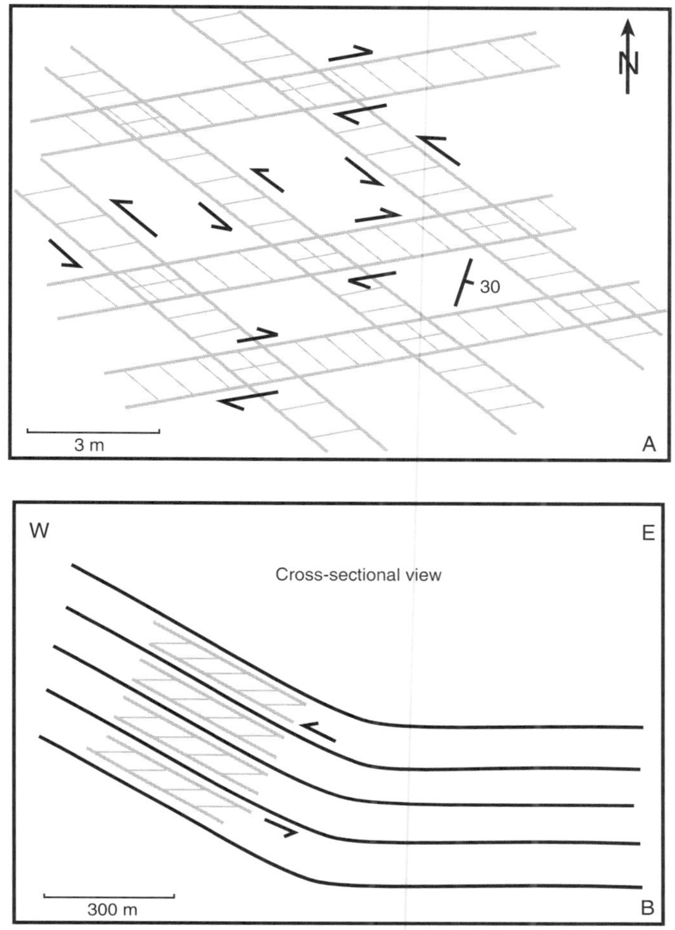

Figure 106. A: Schematic view of pattern of deformation-band shear zones as expressed in plan view in the Cad's Crotch subarea. B: Schematic view of bedding-parallel deformation-band shear zones in the Cad's Crotch subarea.

To the east of the Cad's Crotch subarea, at the Lunch Spring subarea (20 in Fig. 85A) in Governor's Wash, outcrops display perfectly flat-lying Straight Cliffs Formation. There are units within the Straight Cliffs Formation at this location that are full of cataclastic (DB1b) conjugate-thrust deformation bands. They strike northward, approximately parallel to the trend of the East Kaibab monocline. They dip both to the east and west at low to moderate angles and appear to have formed in response to compressive stress nearly perpendicular to the direction of the East Kaibab monocline. Perhaps these are of the DB3 variety, i.e., the "compactional bands" of Mollema and Antonellini (1996). The Strangelove subarea (28 in Fig. 85A) occupies a comparable structural position, though some 30 km (18 mi) to the south. The deformation bands at this locality strike essentially parallel to the trend of the East Kaibab monocline, but they dip steeply and are of undetermined offset.

Analysis of deformation bands in Navajo Sandstone in the southernmost part of the Kaibab subregion

Reconnaissance structural analysis of deformation bands and zones of deformation bands was carried out along the East Kaibab monocline well to the south in the vicinity of Highway 89, and along the dirt road that branches south from Highway 89 along the monocline (Fig. 2C). The main localities are the West Sand Cove subarea (29 in Fig. 85A) and the Power Line subarea (30 in Fig. 85A). Bedding in Navajo Sandstone strikes north-northeast and dips to the southeast at both locations. DB1c ladder-structure deformation-band shear zones abound, as do slickenlines on polished fault surfaces. The patterns and fabrics reflect an oblique-slip deformation, perhaps once again reflecting the fact that the east margin of the Kaibab uplift was marked in Laramide time by significant right-handed strike-slip movement. At West Sand Cove (29 in Fig. 85A), the very top of the Navajo Sandstone is exposed as a stripped bedding surface, striking ~N10°E and dipping ~30°E. Two main sets of deformation bands and zones of deformation bands are apparent: one striking north-northwest, dipping steeply to the northeast, and another striking east-northeast and dipping moderately northwest (29 in Fig. 85B). The former appears to accommodate reverse left-handed shear; the latter appears to accommodate normal right-handed shear. The net effect would appear to be a west-northwest–east-southeast contraction. The Power Line subarea shows very strong systematics (30 in Fig. 85B). There are two prominent sets of deformation-band shear zones, one striking north and dipping steeply east, a second striking northeast and dipping steeply northwest. The northeast-striking set, which is oriented strongly oblique to the strike of bedding, is especially prominently expressed in outcrops (Fig. 107) and accommodates right-handed movement. Thus, as in the case of West Cove subarea, the direction of contraction appears to have been west-northwest–east-northeast.

Origin of the deformation-band shear zones in the subregion

The earliest deformation bands to develop in the Kaibab subregion are pretectonic DB2a structures. Like those in the Zion and Long Valley subregions, these are the noncataclastic deformation bands that occur abundantly in the uppermost Navajo Sandstone.

Figure 107. A: Southwest-directed view showing deformation-band shear zone cutting obliquely across the bedding. Geologists are Tim Tilton and Sarah Tindall. B: Another southwest-directed view showing lenticular deformation-band shear zone crossing the bedding obliquely. Sarah Tindall is the geologist.

Their orientations are consistent with formation by gravitational loading during burial and compaction of the Navajo Sandstone. The deformation bands formed in conjugate-normal systems, bisected by the direction of loading. Within the East Kaibab monocline, these DB2a deformation bands have been rotated out of their original orientations by an amount that relates to the dip of bedding within the Navajo Sandstone. In general, easterly-dipping pretectonic DB2a deformation bands have been rotated to steeper east dips. West-dipping pretectonic DB2a deformation bands have been rotated to shallower west dips. The greatest principal stress (σ_1) for the pretectonic deformation bands was vertical. The deformation involved a collapse of porosity but without cataclastic flow, only some microcracking. Overall the deformation involved a slight volume loss, and thus there was no attendant horizontal extension, except perhaps locally.

The DB2a deformation bands in the Kaibab subregion are clearly cut and offset by tectonic deformation bands of a cataclastic nature that formed during the development of the East Kaibab monocline. The East Kaibab monocline is a fault-propagation fold that developed as a consequence of major reverse right-handed faulting (Tindall and Davis, 1999). The reverse-slip component along the master fault is ~1.6 km (1 mi), and the right-handed strike-slip component is ~4 km (2.5 mi) and could be as great as ~8 km (4.8 mi). The oblique-slip origin is clearly registered in the geometry and kinematics of the deformation band-shear zones.

Several different varieties of the Laramide cataclastic deformation-band shear zones can be distinguished in relationship to the development of the East Kaibab monocline. In the very earliest stages of transpressional deformation, when the Navajo Sandstone was still essentially flat lying, oblique stretching took place in the region that would become the crestal part of the upper hinge of the East Kaibab monocline. This stretching resulted in the formation of the well-developed system of northeast-striking conjugate-normal deformation-band shear zones, which control the fault-fin landscape. These shear zones strike in a direction that is ~20° to 30° clockwise of the axial trace of the East Kaibab monocline, reflecting the influence of the right-handed component of shearing along the master fault at depth. Within this conjugate system of deformation-band shear zones, the individual shear zones are Riedel in nature (Davis et al., 1999). Principal stress directions can be inferred on the basis of the orientations of the faults and slickenlines. The greatest principal stress (σ_1) was oriented N81°E–S81°W and plunged 1°W; the intermediate principal stress (σ_2) was oriented N20°W–S20°E and plunged 48°NW; and the least principal stress (σ_3) was oriented N9°W–S9°E and plunged 41°S.

As the East Kaibab monocline developed at the stratigraphic level of the Navajo Sandstone, the earliest conjugate-normal expression of cataclastic deformation-band shear zones was rotated about the N20°E axis of the fold by an amount related to rotation of bedding from its original horizontal attitude. As a result, originally east-dipping, normal-offset deformation-band shear zones were rotated to steeper attitudes. Some even were rotated into steep west dips, whereupon they became reverse-offset deformation-band shear zones.

The cataclastic deformation-band shear zones in the steep middle limb of the East Kaibab monocline are for the most part top-to-the-east structures, including thrust-slip, reverse-slip, and normal-slip shear zones. Most of the deformation-band shear zones are oblique to bedding and cross-bedding, though some are parallel to bedding and cross-bedding. Details of their development have not yet been worked out. The working hypothesis is broad-based: some of the normal-slip deformation bands reflect a layer-parallel stretching as the Navajo Sandstone was folded; some of the reverse-slip and thrust-slip deformation-band shear zones reflect shortening; and most if not all of the layer-parallel deformation bands accommodated flexural-slip folding. What has not been worked out is *when* each deformation-band shear zone developed relative to the rotation of the Navajo Sandstone bedding during folding. Some early-formed normal-slip deformation-band shear zones were undoubtedly rotated into thrust or reverse appearances. Some early-formed thrust- and reverse-slip deformation-band shear zones were undoubtedly rotated into normal-slip appearances.

The cataclastic deformation-band shear zones in the Straight Cliffs Formation and Wahweap Sandstone give additional kinematic insights. Some of these deformation bands are located adjacent to the lower hinge of the East Kaibab monocline, where the effects of out-of-the-syncline flexural slip are expressed in the layer-parallel, top-to-the-west, deformation bands. Some of the deformation bands are located near the trace of the northernmost expression of reverse right-handed faulting. The effects of this transpressional deformation are expressed in the system of conjugate-oblique shear zones. Furthermore, there are some deformation-band shear zones in flat-lying Straight Cliffs and Wahweap strata immediately east of the lower hinge of the East Kaibab monocline. These formed in conjugate-thrust patterns, with individual deformation bands striking essentially parallel to the trend of the monocline. This geometry suggests that the horizontal contractional stresses associated with the development of the East Kaibab monocline reached eastward beyond the lower hinge of the monocline.

ANALYSIS OF DEFORMATION BANDS, THE KAIPAROWITS SUBREGION

General nature of the subregion

The Kaiparowits subregion lies between the Kaibab uplift and the Circle Cliffs uplift (Figs. 2B and 108A). It includes the folded Upper Cretaceous strata of the Kaiparowits Plateau, the expansive exposures of Navajo Sandstone in the Escalante region, and the Entrada Formation exposures that crop out so abundantly along the Hole-in-the-Rock Road, which connects Escalante with Lake Powell (Fig. 2, B and D). The fold pattern within the Kaiparowits subregion is unusual (Fig. 108A), for in the southern part of the Kaiparowits basin there is a northwestward conver-

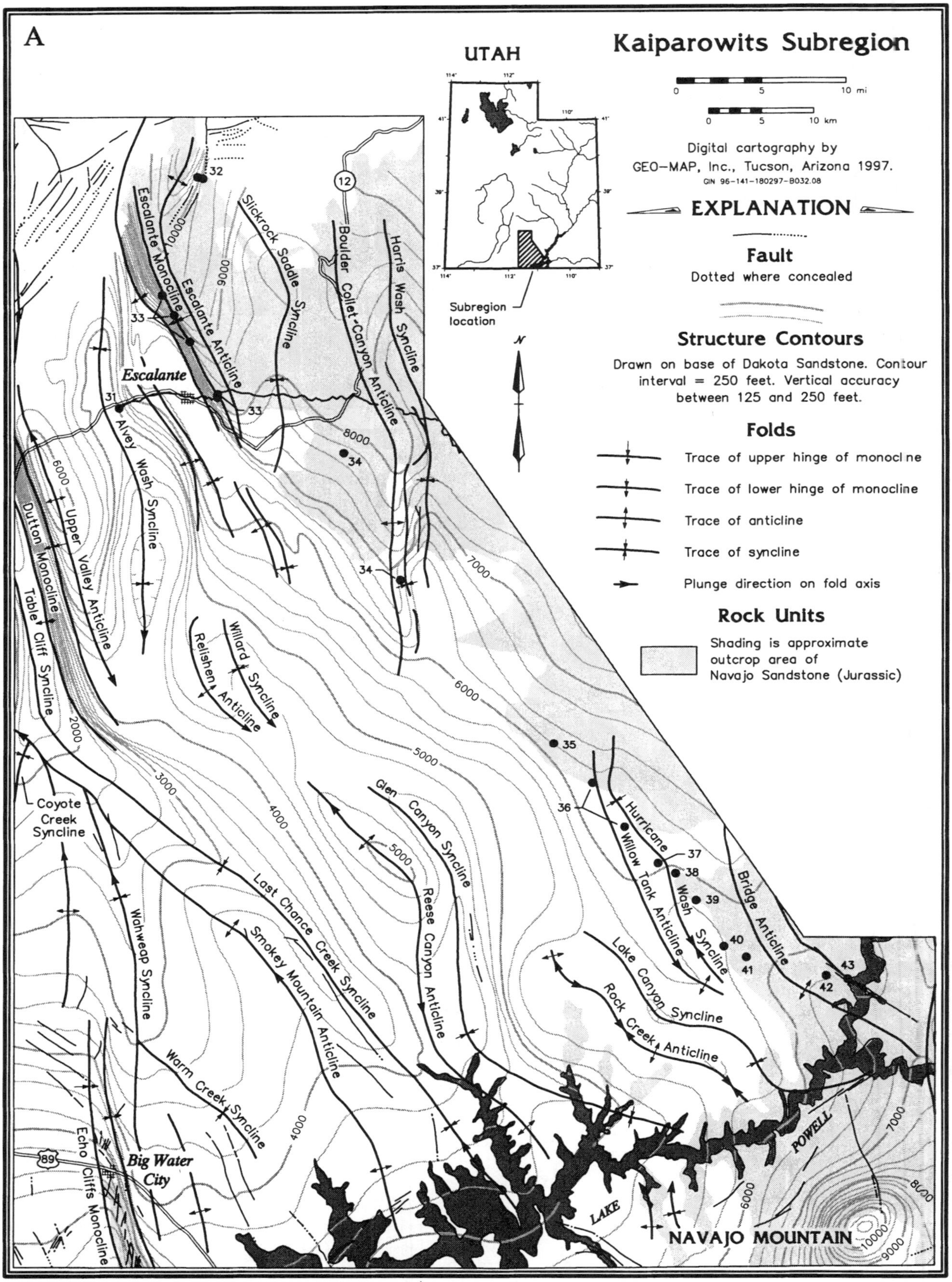

Figure 108 (this and next two pages). A: Map of the Kaiparowits subregion, showing distribution of Navajo Sandstone and locations of subareas where deformation-band data were collected.

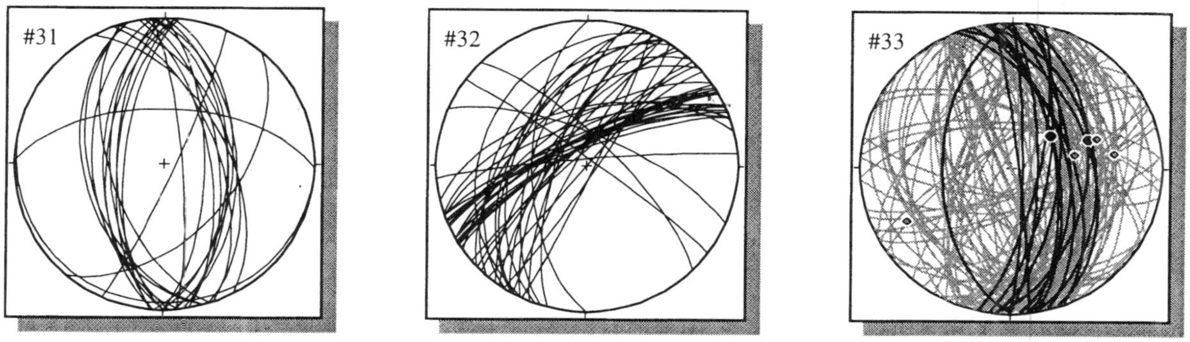

ALVEY SYNCLINE
Straight Cliffs Formation
N = 25

HELL'S BACKBONE
Navajo Sandstone
N = 47

ESCALANTE MONOCLINE
Uppermost Navajo Sandstone
Reverse = 15 Other = 86

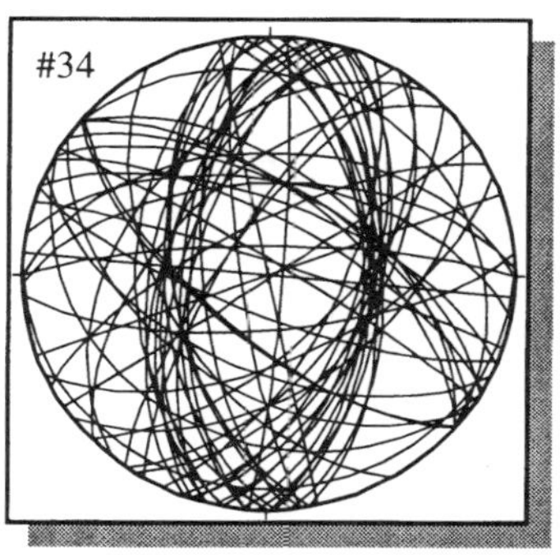

SEEP FLAT
Uppermost Navajo Sandstone
N = 63

B

Figure 108 (this and next page). B: Stereographic projections of deformation bands (great circles) and slickenlines (dots) measured in the Kaiparowits subregion. Numbers 31–43 refer to the subareas located in A. *N* = number of observations.

gence of a number of the subregional anticlines and synclines. The overall northwest trend of the folds in the Kaiparowits basin reflects northeast-southwest Laramide tectonic contraction. The abrupt northward convergence of folds in the western part of the Kaiparowits basin may in fact reflect indention of the western part of this Cretaceous basin by the Kaibab uplift. Transpressive, reverse, right-handed movement of the Kaibab uplift may have "crowded" the strata of the western margin of the Kaiparowits Plateau and created the unusual converging fold pattern.

Another interesting aspect of the folding in the Kaiparowits subregion is that the Dutton and Escalante monoclines are *west*-facing and thus antithetical to the regional eastward-facing signature of the big monoclines, i.e., the East Kaibab, the Waterpocket, and the San Rafael. Presumably the Dutton and Escalante monoclines are each rooted to eastward-dipping, blind, reverse or thrust faults.

There are a variety of deformation bands in the Kaiparowits subregion (Appendix Table 2). Noncataclastic DB2a conjugate-normal deformation bands occur at the very top of the Navajo Sandstone within much of the Kaiparowits subregion. But cataclastic (DB1) deformation bands and zones of deformation bands in the Navajo Sandstone and Straight Cliffs Formation are relatively uncommon throughout this subregion. Extensive reconnaissance examination of sandstones within the Straight Cliffs Formation in the Kaiparowits Plateau turned up very few localities containing deformation bands. Similarly, extensive reconnaissance examination of the Navajo Sandstone within the Escalante and Calf Creek region (Fig. 109) confirmed the *absence* of deformation bands of cataclastic origin. I had originally hoped that there would be a record of tectonically induced deformation banding in the Navajo Sandstone even within regions of essentially horizontal strata and thus a record of stress transmission into and through the Colorado Plateau. My conclusion, however, is that deformation-band shear zones of tectonic

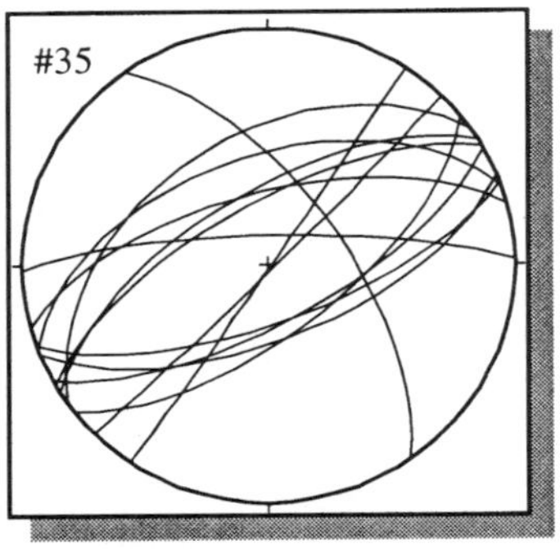

DRY FORK
Uppermost Navajo Sandstone
N = 13

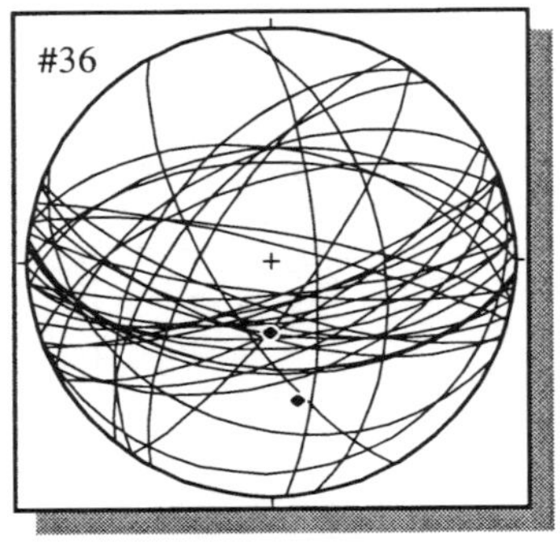

BENCHMARK
Entrada Sandstone
N = 35

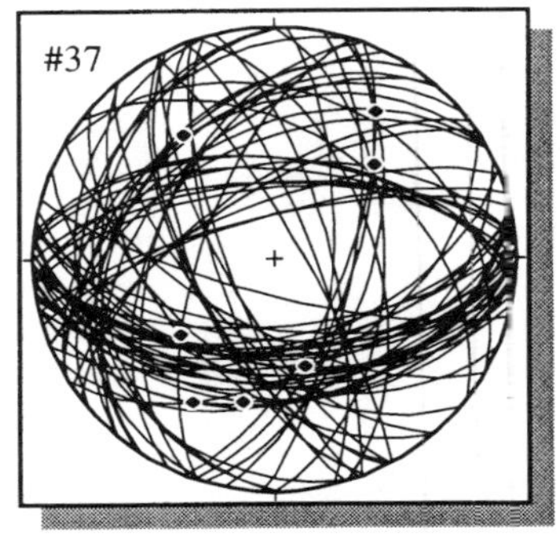

DANCE HALL
Entrada Sandstone
N = 80

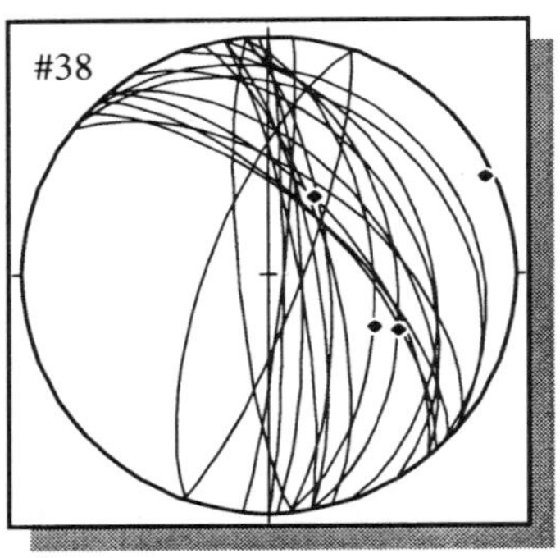

FORTYMILE SPRING
Entrada Sandstone
N = 22

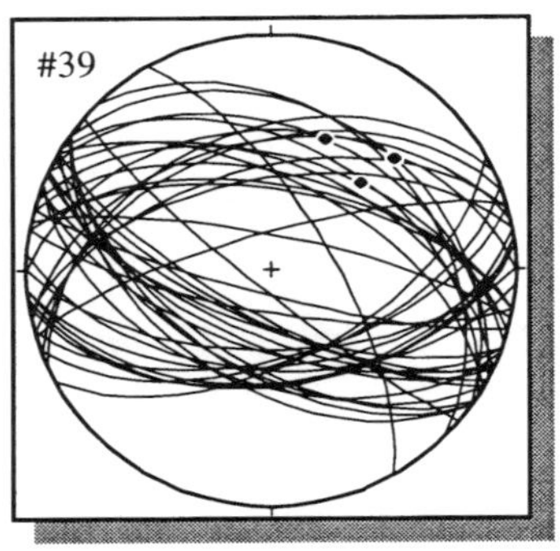

SOONER ROCKS
Entrada Sandstone
N = 43

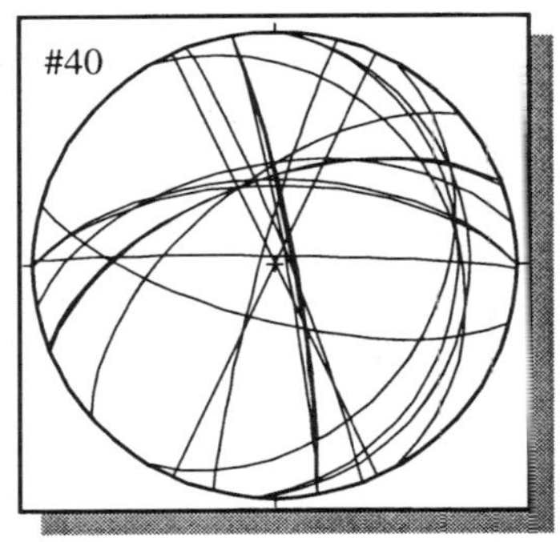

CAVE SPRINGS
Entrada Sandstone
N = 19

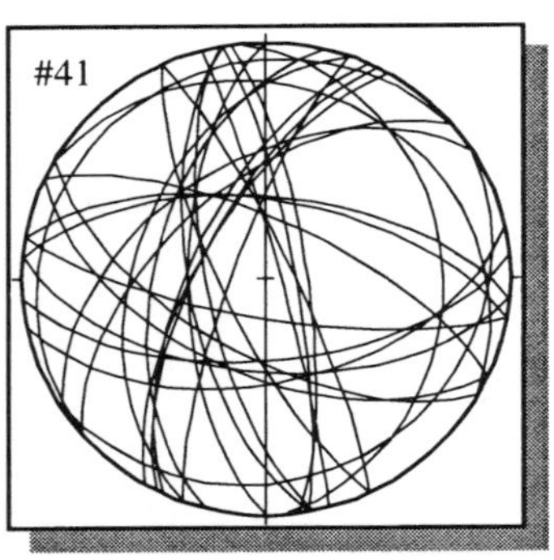

SODA SPRINGS
Entrada Sandstone
N = 30

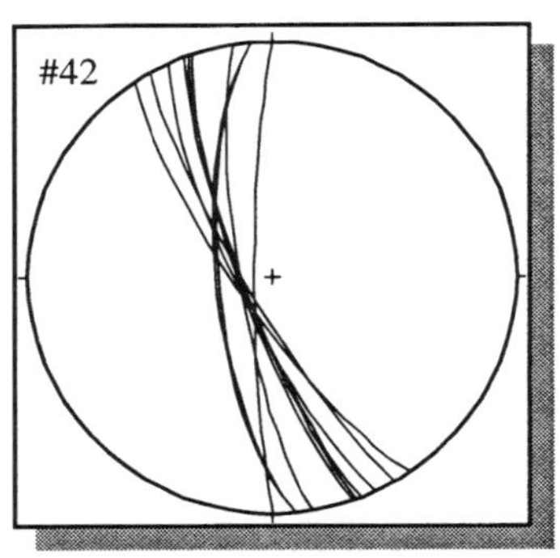

ROCK ROAD
Uppermost Navajo Sandstone
N = 10

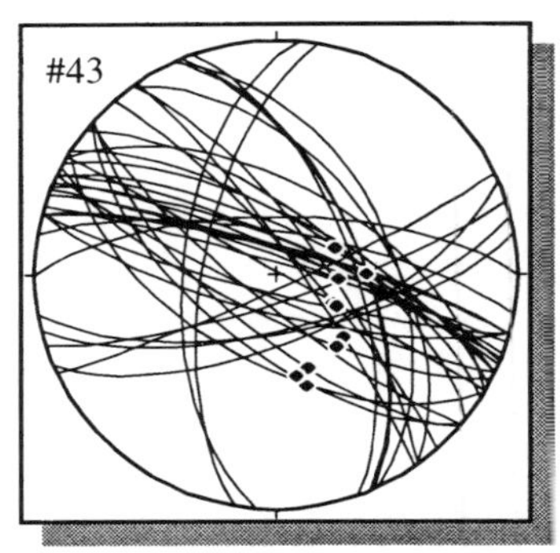

HOLE-IN-THE-ROCK
Navajo Sandstone
N = 40

B (continued)

Figure 109. West-directed view of expansive Navajo Sandstone northeast of Escalante in the Calf Creek area. The team is heading down the trail for a cool swim at Lower Calf Creek Falls.

origin only occur in the Navajo Sandstone where it has been affected by major regional structures.

The record of deformation banding in sandstones of the Entrada Formation is, in contrast, extremely well developed within the Kaiparowits subregion. Indeed, the systematic orientation and kinematics of deformation bands within the Entrada Formation, and the regional pervasiveness of the deformation banding, suggest some regional tectonic control. Yet, interpreting the significance of the deformation banding within sandstones of the Entrada Formation is not without its special challenges.

Properties of deformation bands in Straight Cliffs Formation, northern Kaiparowits subregion

There is from place to place a modest development of deformation bands associated with Laramide anticlines and synclines in the Kaiparowits basin. One of the best and most easily accessible examples occurs in the Straight Cliffs Formation in the hinge zone of the Alvey syncline (31 in Fig. 108A). Sandstone in the hinge of the fold displays conjugate normal zones of deformation bands of a noncataclastic DB2 nature (Fig. 110). There are no slickenlines, slickensides, or ladder structures; and displacements are millimeter scale where seen. The conjugate nature of the orientations of these deformation bands is well displayed stereographically (31 in Fig. 108B). The deformation bands strike ~N10°W and dip at ~60° both to the east and and to the west.

The structural confinement of these zones of deformation bands to the hinge of the Alvey syncline suggests a cause-and-effect relationship. The immediately underlying formation is the mechanically incompetent Tropic Shale. The deformation bands may represent extension and flattening on the outer arc of the fold in the sandstone. The noncataclastic nature of the fabric within these zones of deformation bands might suggest that the deformation mechanism was granular flow and that the sandstone of the Upper Cretaceous Straight Cliffs Formation may not have been completely lithified at the time of this Laramide folding.

Properties of deformation bands in Navajo Sandstone, northern Kaiparowits subregion

More productive with respect to deformation-band development are outcrops along the Escalante monocline (Fig. 108A), a west-facing monocline of beautiful topographic expression (Fig. 41). The Escalante River carves transversely through the structure and exposes deformation bands and zones of deformation bands that are not evident except in the monoclinal limb proper. The most accessible part of the monoclinal limb is uppermost Navajo Sandstone, which is expressed as an erosionally

Figure 110. South-directed photo of conjugate-normal deformation bands of a noncataclastic origin in the hinge of the Alvey syncline. Geologist is Seth Gering.

stripped structural surface. Beds dip ~25°W. The deformation-band pattern appears quite complex, for the uppermost Navajo Sandstone contains abundant, closely spaced noncataclastic DB2a deformation bands. These deformation bands are abundantly developed along the structure and, given the additional overprint by deformation bands of a cataclastic nature, present a bewildering array of structural complexity. The complexity is partly revealed in the stereographic projection of measurements of deformation bands in the Escalante monocline subarea (33 in Fig. 108B) The noncataclastic deformation bands, plotted as gray great circles, are arranged in "ball of string" fashion. The cataclastic deformation bands, plotted as black great circles, strike consistently north, parallel to the trend of the Escalante monocline, and dip moderately eastward. Many of the deformation bands are black, coated with manganese and iron, revealing the degree to which the impermeable deformation-band shear zones influence the circulation of fluids. None of the noncataclastic deformation bands is associated with slickenlines, slickensides, or ladder structures. They cut through slump folds and soft-sediment fault structures in the Navajo Sandstone, but in turn they are cut by true cataclastic, tectonic deformation bands. The superimposed cataclastic deformation bands are, for the most part, relatively thick (up to 10 cm [4 in.]) and commonly are marked by polish and slickenlines. They display reverse offset.

Just beyond the northern tip of the Escalante monocline, where Navajo Sandstone is essentially flat lying, there is a wonderful illustration that deformation bands of a cataclastic nature simply are produced in the Navajo Sandstone *only* in proximity to major regional structures. The locality is the Hell's Backbone subarea (32 in Fig. 108A), ~20 km (12 mi) north of Escalante. As the name "Hell's Backbone" implies, this is an area of remarkably rugged landscape, fashioned from Navajo Sandstone (Fig. 42). Well to the east and east-southeast of this subarea, the Navajo Sandstone is absolutely devoid of deformation bands, yet they "just suddenly show up" at Hell's Backbone. The presence of these bands alerted my students and me to the presence of geologic structure. Geologic maps of the region reveal the presence of several large, throughgoing, N30°–45°E-striking faults that pass through the Hell's Backbone area (Fig. 2, B and D). The faulting appears to be Basin and Range in origin, for it heads north and cuts Miocene volcanic rocks of the Aquarius Plateau. The deformation-band shear zones at Hell's Backbone tend to be relatively thick (up to 10 cm [4 in.]). Offsets along the deformation-band shear zones are normal, where they can be discerned. Stereographic projection of the measurements collected in the Hell's Backbone subarea reveal that the cataclastic deformation bands range in strike from ~N15°E to N65°E and thus are subparallel to the major faults that project through Hell's Backbone (32 in Fig. 108B). Dip direction of the deformation-band shear zones is northwest, at moderate to steep angles.

Properties of deformation bands in Navajo Sandstone, southern Kaiparowits subregion

Hole-in-the-Rock Road leads from Escalante south-southeastward to Lake Powell, and it follows the base of the Straight Cliffs (Fiftymile Mountain) escarpment, which lies to the west (Fig. 108A). Along this stretch there are outcrops of essentially flat-lying Mesozoic strata, notably uppermost Navajo Sandstone, Carmel Formation, and Entrada Formation. The Navajo Sandstone and sandstones of the Entrada Formation carry deformation bands, and these were studied and measured as a part of the project. With the exception of what was found in just one subarea (Hole-in-the-Rock, 43 in Fig. 108B), the deformation bands in the Navajo Sandstone appear to be largely of noncataclastic origin.

Uppermost Navajo Sandstone is exposed in the Seep Flat (subarea 34), Dry Fork (subarea 35), Rock Road (subarea 42), and Hole-in-the-Rock subareas (Fig. 108A). Deformation bands in these subareas all are devoid of slickenlines and slickensides. Offsets, where discernible, are almost all tiny normal displacements. The predominant strike of the deformation bands at the Seep Flat subarea (34 in Fig. 108B) is north, with conjugate 60° dips to east and west. In stereographic projection, these sets stand out from a background of deformation bands with a broad range of strike, dip, and dip direction (34 in Fig. 108B). In the Dry Fork subarea

(35 in Fig. 108A), the predominant deformation bands are conjugate normal, striking east-northeast, with the bands dipping ~60° both to the southeast and the northwest (Fig. 108B). The record of deformation bands at Rock Road (42 in Fig. 108B) is very feeble; approximately eight thin deformation bands were measured, which strike north-northwest and dip steeply to the southwest.

In contrast to these three subareas where only uppermost Navajo Sandstone is exposed and where there are no major local structures, the Hole-in-the-Rock subarea (43 in Fig. 108A) samples much of the full thickness of the Navajo Sandstone along the precipitous Hole-in-the-Rock trail that leads down to Lake Powell (Fig. 111). The historically famous Hole-in-the-Rock marks the location where Mormon pioneer settlers achieved the impossible, by fashioning a trail along which they could lower their wagons and provisions down the ~350+ m (1,155 ft) cliff, thus providing access to the Colorado River bottom, which they would cross on the way to settling what is now Bluff, Utah. The trail follows a large normal fault that strikes ~N55°W and dips steeply west (Fig. 108A). Offset appears to be a few 10s of meters. Quite a number of deformation bands and zones of deformation bands can be seen along the steep Hole-in-the-Rock trail, and at various depths within the Navajo Sandstone. Slickenlines are evident from place to place, and these were measured. The greatest observed displacement on any of the zones of deformation bands was found to be about 10 cm (4 in.). Like the fault, the deformation bands strike ~N55°W (43 in Fig. 108B). The stereographic pattern of the deformation bands is one of conjugate-normal faulting, with sets dipping moderately to steeply both to the northeast and the southwest. Those dipping northeast have a modest right-lateral component; those dipping southwest have a modest left-lateral component (Fig. 108B).

Figure 111. South-directed photograph of the head of Hole-in-the-Rock with Lake Powell in the background. Geologist is Tim Tilton.

Properties of deformation bands in Entrada Formation, southern Kaiparowits subregion

Within the Kaiparowits subregion, deformation bands and zones of deformation bands are well developed in the Entrada Formation, occurring in the form of pink and orange to red and orange, generally nonresistant, noncataclastic structures accommodating very modest normal displacements. The outcrop expression of the deformation bands in sandstones of the Entrada Formation is distinctly different from deformation bands within the Navajo Sandstone and Straight Cliffs Formation. The deformation bands within the Entrada Formation do not tend to stand out in topographic relief, but instead appear to have the same resistance to weathering and erosion as the bedrock host (Fig. 112). The deformation bands in the Entrada Formation are not at all planar, but instead are curviplanar and braided in outcrop-trace expression. Outcrop traces of the DB2c subvariety (see Appendix Table 1) typically are irregular and anastomosing, with spacings commonly on the order of 1 to 3 m (see Fig. 112). The orientations are quite systematic in any given outcrop, less systematic from one subarea to another, though with sufficient tendency toward preferred orientation that it is necessary to consider the possibility of regionally systematic patterns.

Deformation bands in the Entrada Formation within the Kaiparowits subregion tend to strike east to east-northeast, or north to north-northwest, admittedly with exceptions. All appear to be conjugate normal, with dip inclinations typically ranging from ~40° to 70°. Photomicrographs of deformation bands in the Entrada Formation reveal loss of porosity through pore-space collapse. Grain fracturing is relatively common, but cataclastic flow is rare.

At the Benchmark subarea (36 in Fig. 108A), the deformation bands within the Entrada Formation are quite plentiful, but very thin, on the order of millimeters. They belong to the DB2d subvariety (see Appendix Table 1). Offsets are normal slip; most are on the millimeter scale, but some are on the centimeter scale. The dominant set of deformation bands at the Benchmark subarea strikes east and dips moderately to steeply southward (36 in Fig. 108B). Slickenlines are rare, but two were noted, and these suggested dip-slip movement. A less well developed conjugate-normal set dips northward. In addition, there is a minor population of deformation bands that strike north-northwest and dip steeply, both east and west.

Figure 112. Outcrop expression of irregular, anastomosing, DB2c nonresistant deformation bands that are so common within the Entrada Formation in the Kaiparowits subregion.

Figure 113. West-directed photograph of Dance Hall Rock. Bedrock is massive pink sandstone of Entrada Formation. Justin Banz is waving from the top, and just below him is what I interpret to be a fossil plunge pool, formed by a falls that once came over the top before the source was pirated. In the foreground are light pink traces of anastomosing DB2c deformation-band shear zones.

The most spectacular locale for deformation bands in the Entrada Formation is Dance Hall Rock (37 in Fig. 108A). Dance Hall Rock is a smooth, red, slickrock exposure of Entrada Sandstone that served as an amphitheater for Mormon pioneer settlers (Fig. 113), where they gathered together for fun to break the work and tension embodied in the extraordinary effort expended in constructing the Hole-in-the-Rock Road. Somewhat wavy and inosculating to anastomosing in appearance, the zones of DB2c deformation bands can be traced for 10s of meters across the slickrock landscape (Fig. 114). Tape-and-compass mapping was carried out to create a structural geologic map of the details of deformation banding (Fig. 115). The map pattern of these is anastomosing and inosculating. Transverse to the main direction are short, discontinuous deformation bands. None of the zones of deformation bands stand in topographic relief, but rather have the same resistance to erosion as that of the Entrada Formation host rock. Offsets all appear to be normal, though there are no slickensides or slickenlines. Cross-sectional views at Dance Hall Rock reveal that the deformation bands dip moderately to steeply and confirm an accommodation to modest or minuscule normal-slip movements. There are some right- and left-lateral separations of offset features, such as other zones of deformation bands and cross-bedding, but these lateral offsets are probably simply separations resulting from the normal displacement of inclined features. Microscopic examination reveals that these bands are noncataclastic, and perhaps this character helps account for the fact that they are not particularly resistant to erosion. Reduction spots occur within the pink Entrada sandstone, and where the zones of deformation bands cross areas of reduction-spotted Entrada rocks, the reduction spots are not distorted.

The dominant sets of deformation bands at Dance Hall Rock strike east and dip moderately to the north and to the south (37 in Fig. 108B). These sets are essentially identical to those measured in the Entrada Formation at the Benchmark subarea (compare plots for subareas 36 and 37 in Fig. 108B). Yet another conjugate-normal system of deformation bands at Dance Hall Rock strikes north-northwest, with sets dipping at ~60° both to the northeast and to the southwest (37 in Fig. 108B). Again, the pattern is similar to the subordinate pattern measured in the Benchmark subarea (36). Taken together, the deformation-band system at the Benchmark and Dance Hall subareas appears to consist of two

Figure 114. Outcrop expression of single DB2c deformation-band shear zone in Entrada Sandstone at Dance Hall Rock. Note clipboard for scale.

orthogonal conjugate-normal systems, one striking east-northeast, the other striking north-northwest.

Although subject to some interpretation, the Dance Hall Rock pattern of deformation bands appears to hold within the Fortymile Spring subarea (38) and the Sooner Rocks subarea (39) (Fig. 108B). The Fortymile Spring subarea lies just 1 km (0.6 mi) southeast of Dance Hall Rock, and there the deformation bands are quite pronounced in outcrop (Fig. 116). They display normal offsets, but again with very small displacements. The average strike of the deformation bands is north-northwest; their dips to the northeast range from gentle to steep.

At Sooner Rocks (Fig. 117), the deformation bands are quite similar in physical appearance to those at Dance Hall Rock, ranging from thin and discrete DB2d deformation bands to broad and braided DB2c deformation bands. Normal displacements were recognized in a few places, but the offsets were found to be quite small. Stereographically, the measured orientations of deformation bands at Sooner Rocks are seen to be very systematic, forming a conjugate-normal system that strikes west-northwest and dips both to the northeast and to the southwest (39 in Fig. 108B).

Although deformation bands in the Entrada Formation were measured at yet two additional subareas within the southern part of the Kaiparowits subregion, the patterns are not particularly revealing. At the Cave Springs subarea (40 in Fig. 108B), the deformation bands represent the very thin DB2d subvariety. Displacements are normal and very small (several millimeters), although two deformation bands display reverse offset. No slickenlined or slickensided surfaces were observed. Similarly, at the Soda Springs subarea (41 in Fig. 108B), the bands are very thin, and the displacements are very small and normal in offset. The stereographic projections of the Cave Springs and Soda Springs subarea measurements (40 and 41 in Fig. 108B) again reveal the presence of north-northwest– and east-northeast–striking systems, but there is considerable scatter.

Origin of the deformation-band shear zones in the subregion

The net three-dimensional strain effect of offsets within these systems of deformation bands in the Entrada Formation is vertical shortening, north-northwest–south-southeast stretching, and very slight east-northeast–west-southwest stretching. The stress field responsible for the formation of the deformation bands may have been marked by a vertical greatest principal stress direction (σ_1), a subhorizontal intermediate principal stress direction (σ_2) oriented N85°E–S85°W, and a subhorizontal least principal stress (σ_3) oriented N5°W–S5°E. The directions of intermediate principal stress (σ_2) and least principal stress (σ_3) may have switched at times, creating the two systems of conjugate-normal deformation bands that are seen stereographically (Fig. 108B). Establishing the timing of this deformation has proved to be impossible. The deformation banding in the Entrada Formation may have been generated by gravitational loading during burial and compaction of the sandstone of the Entrada Formation, perhaps within a subtle regional compressive-stress field marked by nearly east-west compression. Alternatively, the deformation banding in the Entrada Formation may have been a response to tectonic loading during the Laramide, in which the greatest principal stress (σ_1) due to depth of overburden was vertical and the intermediate principal stress (σ_2) was compressive and oriented nearly east-west. Because deformation bands within the Entrada Formation are found as well in the Waterpocket, San Rafael, and Henry Mountains subregions, the opportunity arises to reexamine the regional systematics of geometry and kinematics of this family of deformation bands and to reconsider their origin.

ANALYSIS OF DEFORMATION BANDS, WATERPOCKET SUBREGION

General nature of the subregion

The Waterpocket subregion almost exactly corresponds with the full length of the Capitol Reef National Park and includes part of the Lake Powell Recreation Area (Fig. 118A) (Billingsley et al., 1987). I focused most of the field work in this subregion on Navajo Sandstone within the Waterpocket Fold

Figure 115. Tape-and-compass map of the deformation-band shear zone pattern at Dance Hall Rock.

Figure 116. Photograph of DB2c deformation bands in sandstone of the Entrada Formation at Fortymile Spring. Geologist is Sarah Tindall.

proper. However, some data were collected as well in Jurassic and Cretaceous strata lying above the Navajo Sandstone and exposed east of the lower hinge of the Waterpocket Fold. The southern limit of the Waterpocket subregion is Lake Powell, whose shoreline bedrock exposures of Mesozoic strata offer special opportunities for reconnaissance and measurement of deformation bands and zones of deformation bands (Fig. 119). Both the lake and road access into the southern part of the Waterpocket subregion permitted sampling of deformation bands well east of the Waterpocket Fold.

There are several varieties of deformation bands and zones of deformation bands within the Waterpocket subregion (Appendix Table 2). Even this far east in the overall region of study, the upper 10 to 15 m (33 to 50 ft) of Navajo Sandstone continues to display noncataclastic DB2a conjugate-normal deformation bands in abundance. Furthermore, the Navajo Sandstone contains abundant DB1 cataclastic deformation bands and zones of deformation bands, especially right along the Waterpocket Fold. Deformation bands within the Waterpocket Fold subregion also occur in abundance in essentially flat-lying Entrada Formation just to the east of the monocline. Like those in the Kaiparowits subregion, they are mainly thin, pinkish-orange, nonresistant, braided deformation bands of a noncataclastic nature.

Properties of deformation bands exposed along the shores of Lake Powell

Using a rented powerboat (Fig. 119), my students and I (see Acknowledgments) carried out reconnaissance exploration for deformation bands and zones of deformation bands along the bedrock shores of Lake Powell, from Bullfrog Marina (subarea 45 in Fig. 118A) southwest to a location fully 5 km (3 mi) west of the Waterpocket Fold (Fig. 120). We were disappointed in discovering how few deformation bands occur in the Entrada Formation

Figure 117. Southwest-directed photograph of the Sooner Rocks area, revealing the outcrop expression of massive sandstones within the Entrada Formation.

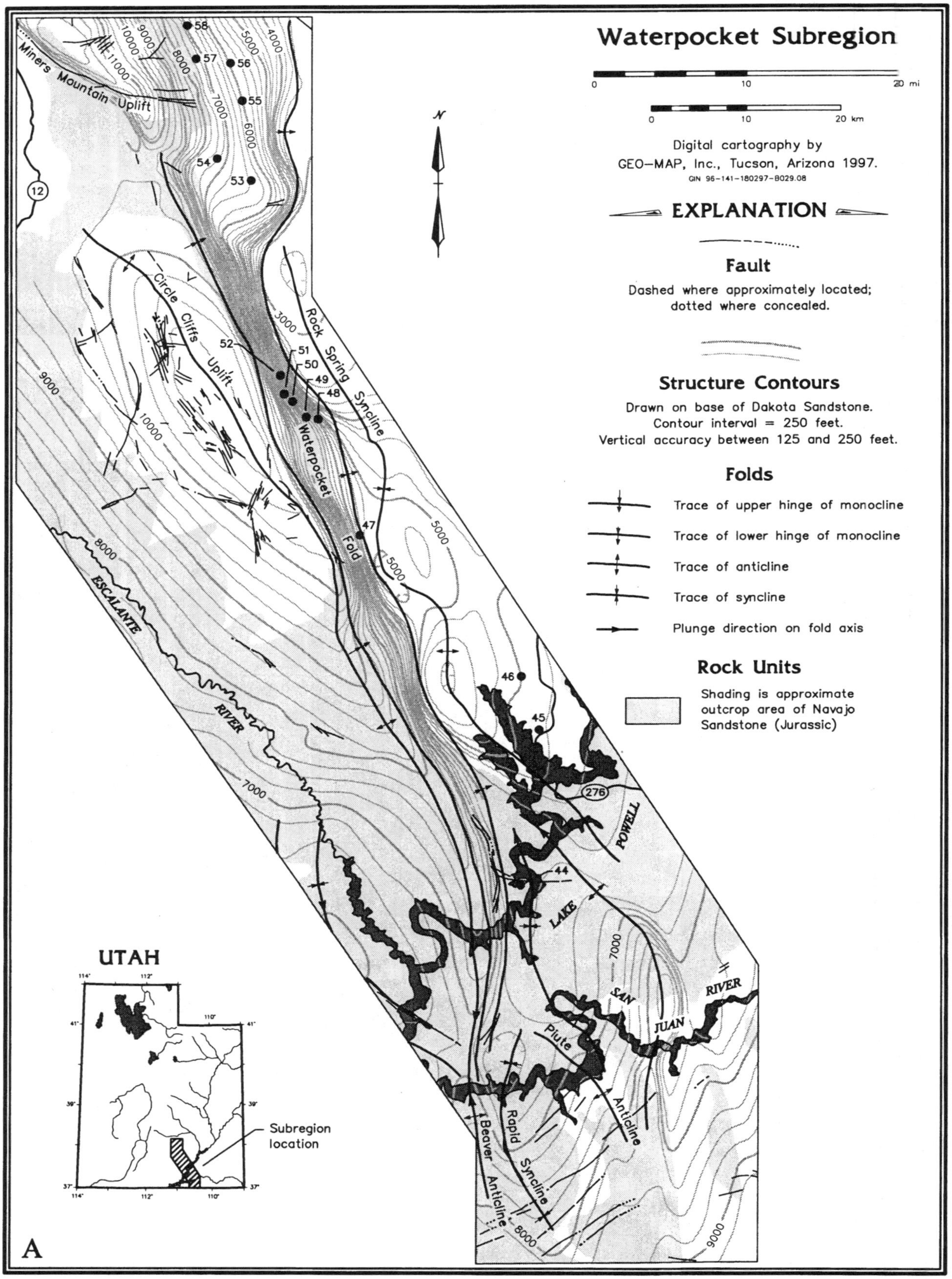
Waterpocket Subregion
0 10 20 mi
0 10 20 km
Digital cartography by
GEO–MAP, Inc., Tucson, Arizona 1997.
GIN 96-141-180297-8029.08
EXPLANATION
Fault
Dashed where approximately located; dotted where concealed.
Structure Contours
Drawn on base of Dakota Sandstone.
Contour interval = 250 feet.
Vertical accuracy between 125 and 250 feet.
Folds
Trace of upper hinge of monocline
Trace of lower hinge of monocline
Trace of anticline
Trace of syncline
Plunge direction on fold axis
Rock Units
Shading is approximate outcrop area of Navajo Sandstone (Jurassic)
Miners Mountain Uplift
Circle Cliffs Uplift
Rock Spring Syncline
Waterpocket Fold
ESCALANTE RIVER
LAKE POWELL
SAN JUAN RIVER
Piute Anticline
Rapid Syncline
Beaver Anticline
UTAH
Subregion location
A

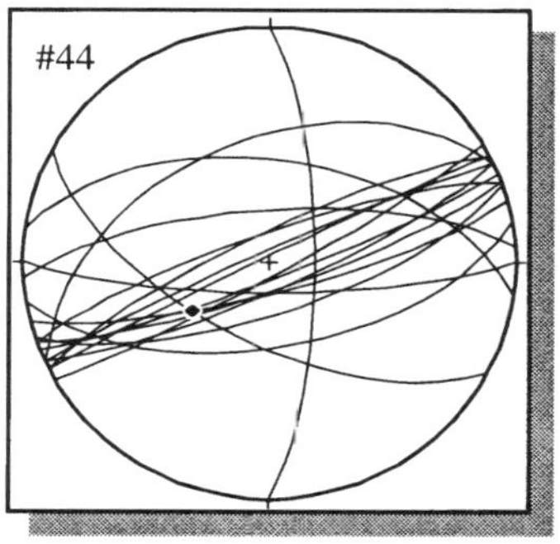

ANNIE'S CANYON
Navajo Sandstone
N = 17

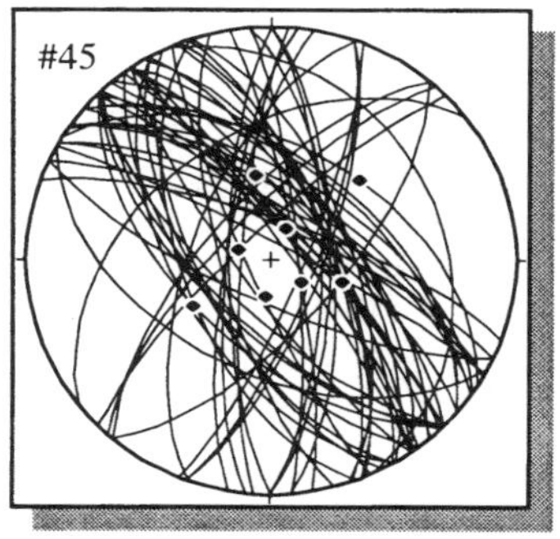

BULLFROG
Entrada Sandstone
N = 63

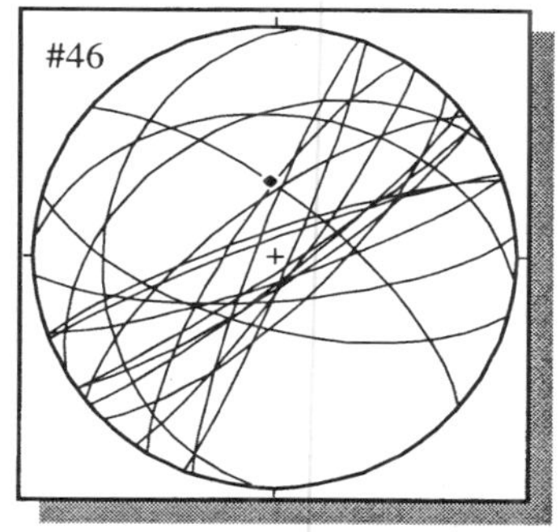

AMY'S
Entrada Sandstone
N = 18

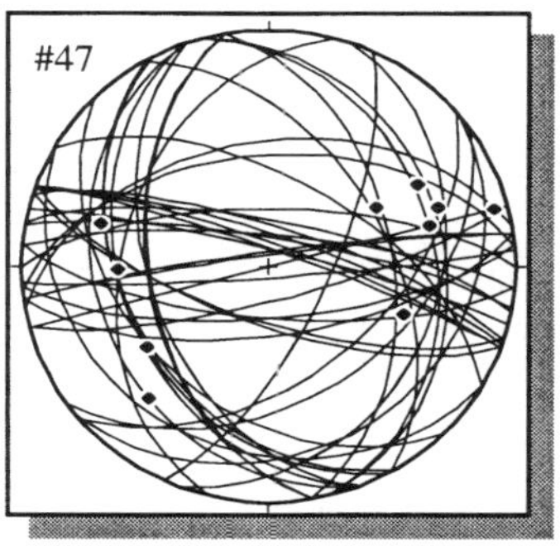

BRIMHALL BRIDGE
Entrada Sandstone
N = 41

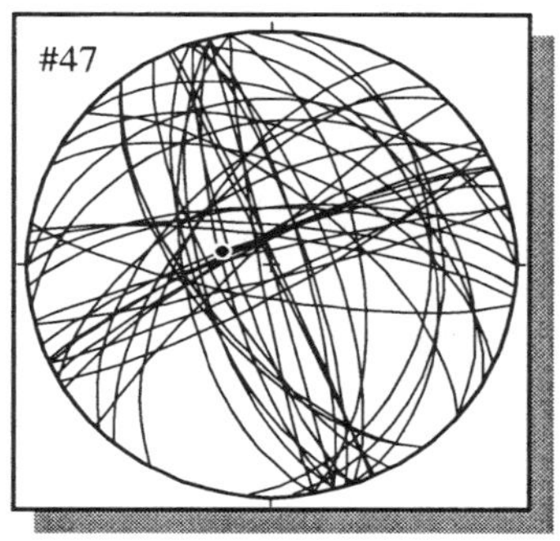

BRIMHALL BRIDGE
Navajo Sandstone
N = 44

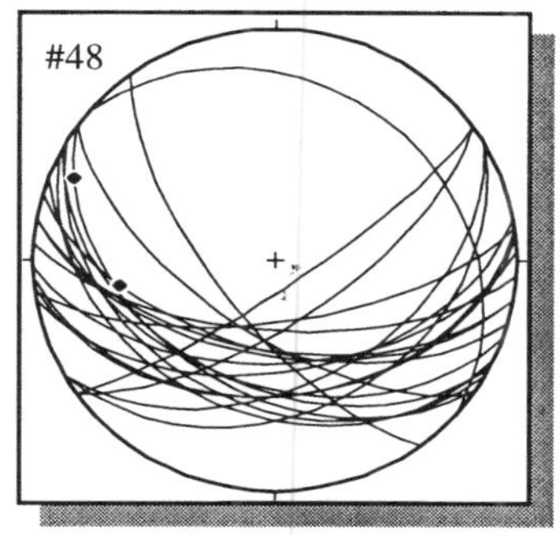

THE POST
Entrada Sandstone
N = 21

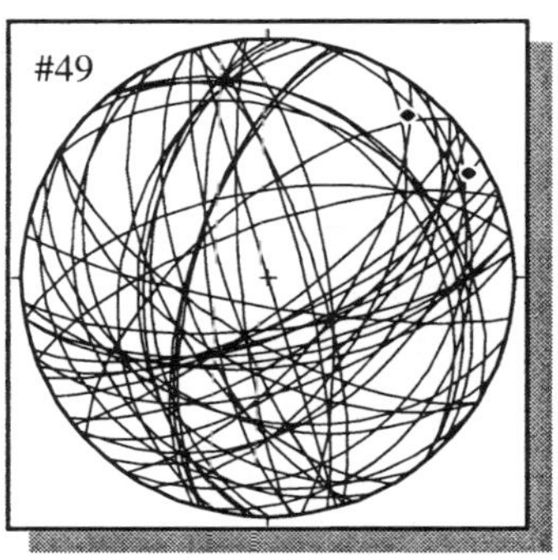

SURPRISE CANYON
Uppermost Navajo Sandstone
N = 50

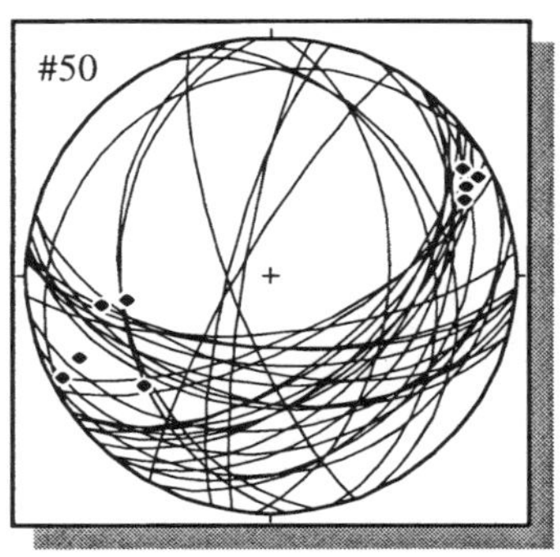

BURR TRAIL
(Undifferentiated)
Navajo Sandstone
N = 46

B

Figure 118 (including previous and next pages). A: Map of the Waterpocket subregion, showing distribution of Navajo Sandstone and locations of subareas where deformation-band data were collected. B: Stereographic projections of deformation bands (great circles) and slickenlines (dots) measured in the Waterpocket subregion. Numbers 44–58 refer to the subareas located in A. *N* = number of observations.

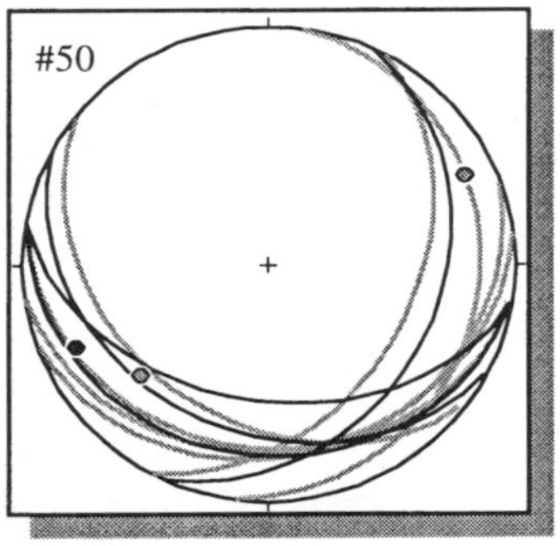

BURR TRAIL
Navajo Sandstone
Normal = 6 Thrust = 4

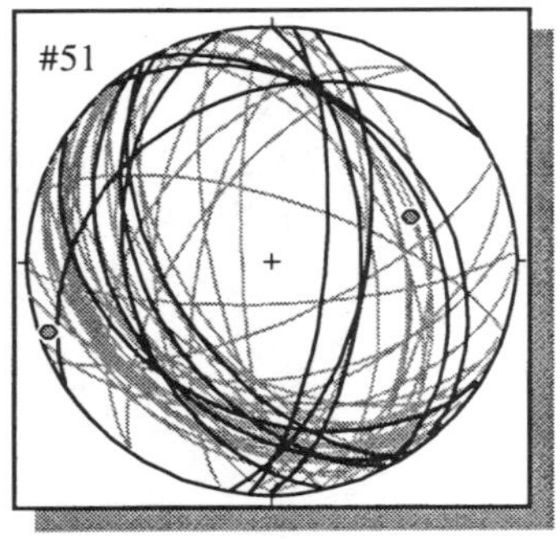

PEEK-A-BOO
Navajo Sandstone
Uppermost Navajo (black) = 12
Others (gray) = 40

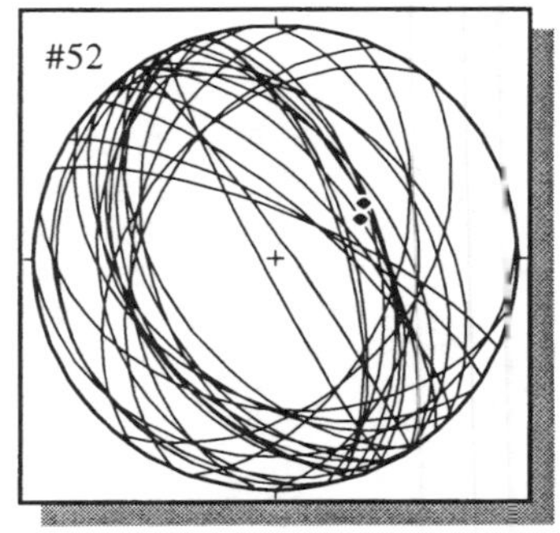

LIZARD CANYON
Entrada Sandstone
N = 36

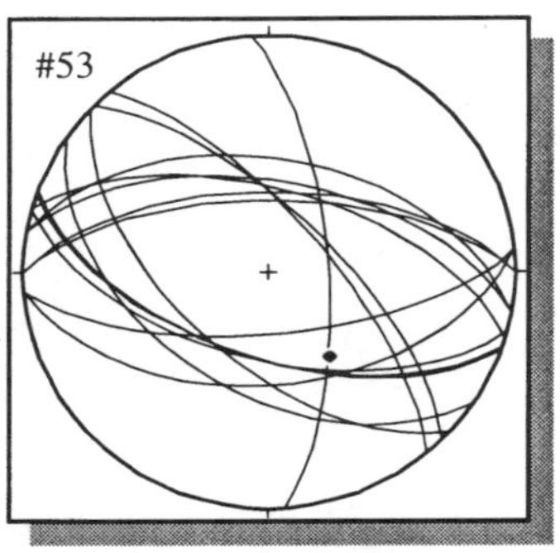

DOGWATER
Entrada Sandstone
N = 15

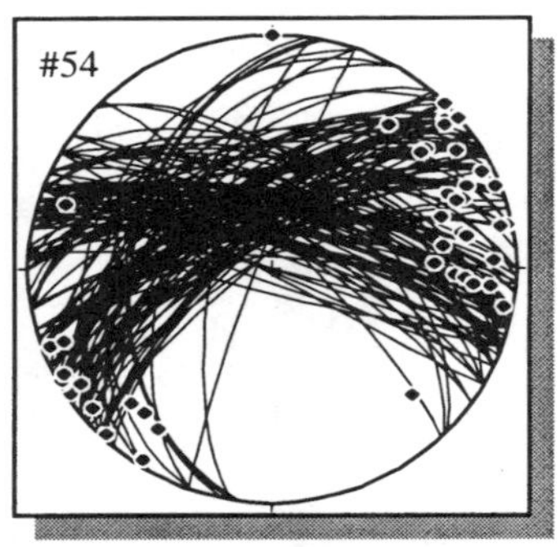

SHEETS GULCH
Navajo Sandstone
N = 156

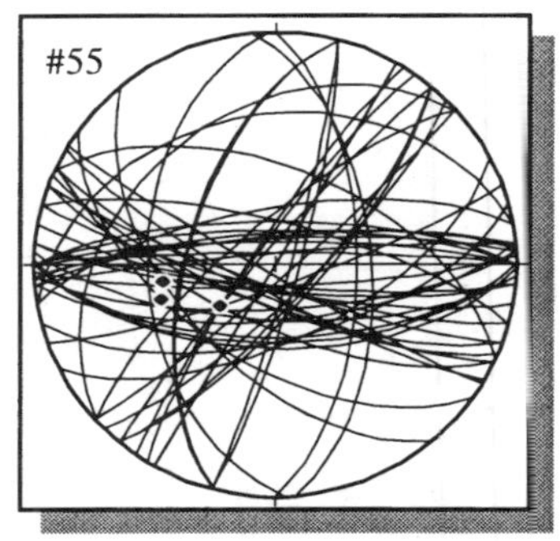

FIVEMILE WASH
Entrada Sandstone
N = 54

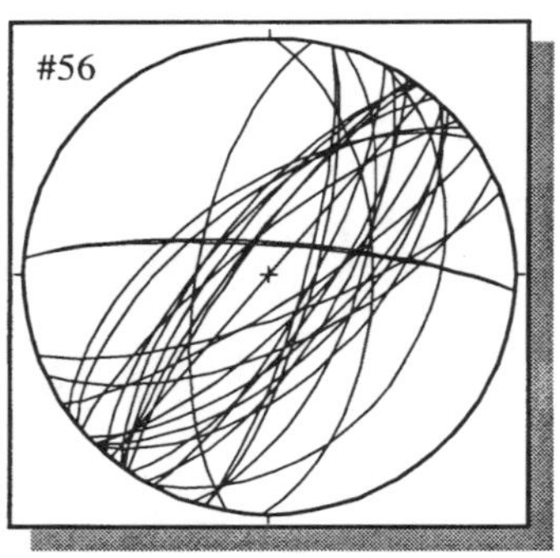

NOTOM
Entrada Sandstone
N = 27

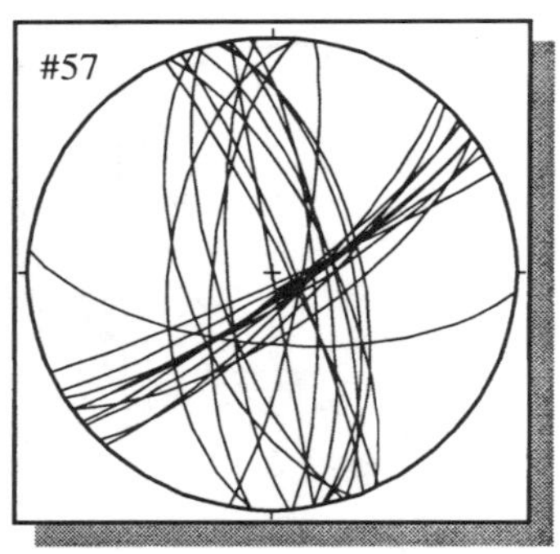

PARADISE CREEK
Uppermost Navajo Sandstone
N = 26

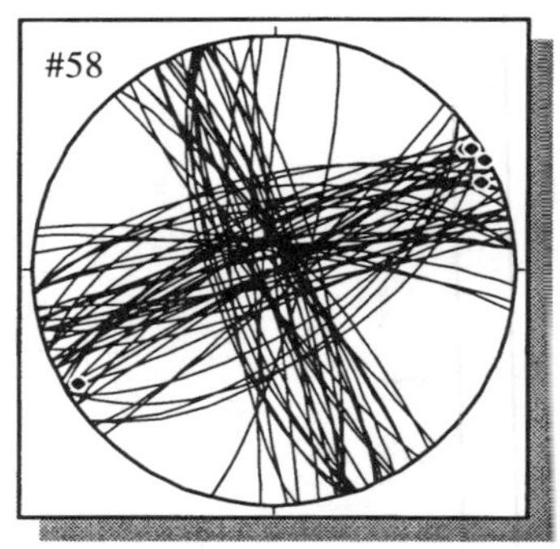

CAPITOL WASH
Navajo Sandstone
N = 75

B (continued)

Figure 119. One of our field vehicles, parked along the shore of Lake Powell. We used this vehicle to explore for deformation-band shear zones in Navajo Sandstone (white, at lake level) and overlying Entrada Formation in the southernmost part of the Waterpocket subregion.

Figure 120. North-directed photograph of gently dipping middle-limb strata of the Waterpocket Fold, near its southern terminus. Lake Powell is in the foreground.

and Navajo Sandstone along the shoreline, where the bedrock is mainly flat lying. And yet this reality helped establish the conclusion that there is essentially no deformation-band record of the propagation of stresses *through* horizontal sandstone bedrock between major regional structures. From place to place, there are deformation bands and zones of deformation bands in the Entrada Formation, but they do not appear to be particularly systematic in organization, nor impressive or interesting in geometry. They are mainly of the DB2d subvariety (see Appendix Table 1). Offsets are negligible.

Only at one location, i.e., Annie's Canyon subarea (44 in Fig. 118A), did we find significant zones of deformation bands. These occur in the Navajo Sandstone well beneath its upper contact with Carmel Formation. The deformation bands, cataclastic in nature, are found along the trace of a conspicuous normal fault (10 m [33 ft] offset) that strikes northeast and dips steeply northwest. The fault shows up clearly on the Utah state geologic map (Hintze, 1980), expressed as a splay that heads east from a fault that parallels the trend of the Waterpocket Fold (Fig. 118A). The zones of deformation bands within Annie's Canyon subarea bear a clear geometric relationship to this fault. Slickenlines were evident along one zone of deformation bands, and they were found to be steeply raking, although the magnitude of offset was not discernible. The deformation bands in Annie's Canyon subarea strike ~N75°E and dip steeply both southeast and northwest. They formed as an accommodation to a modest stretching associated with the formation of the nearby master fault.

In the Bullfrog subarea (45 in Fig. 118A), which is located in the vicinity of Bullfrog Marina on the north shore of Lake Powell, the Entrada Formation is exposed as slickrock over broad, continuous tracts, both along the shore and inland. Deformation bands and zones of deformation bands are abundant, and many can be traced 10s of meters. They are of the DB2c subvariety (see Appendix Table 1). Where a given zone of deformation bands dies out, there is commonly a relay connection with yet another zone of deformation bands, creating either right-stepping and left-stepping arrangements or crudely conjugate relationships (Fig. 121). The zones of deformation bands tend to be anastomosing and inosculating, in places reasonably organized into systems, in other places quite disorganized. Many of the slickrock exposures of the Entrada Formation have scant topographic relief, and thus there is limited opportunity to inspect cross-sectional views for offset relationships. Thus, most separations appear as either right-lateral or left-lateral, even though it is probable that the strike-slip separations were produced by normal-slip movement of gently inclined stratification or cross-stratification. Slickenlines were identified on eight different deformation bands, and in each case the slickenlines were found to rake very steeply. Stereographic display of the measurements of deformation bands and zones of deformation bands in the Bullfrog subarea (45 in Fig. 118B) shows a reasonably consistent pattern, with the dominant conjugate-normal system striking northwest and dipping moderately to steeply both northeast and southwest. In addition there is a much less well developed system that strikes northeast, with sets dipping both northwest and southeast. The dominant northwest strike is parallel to the trace of a long continuous anticline that lies ~2 km (1.3 mi) to the west and on whose northeast limb these deformation bands were measured.

There is a nice exposure of the Entrada Formation in Amy's subarea (46 in Fig. 118A). Bedding is essentially flat lying, cut by noncataclastic deformation bands and zones of deformation bands as thick as several centimeters. Offsets are normal, ranging from about 0.5 to 15 cm (0.2 to 6 in.). Only one slickenlined surface was observed, and it displayed steeply raking lines. Orientations of the bands are somewhat scattered, though the dominant strike is northeast (46 in Fig. 118B).

Figure 121. Outcrop expression of branching DB2c noncataclastic deformation bands in Entrada Formation in the Bullfrog subarea of the Waterpocket subregion.

Properties of deformation bands along the southern half of the Waterpocket Fold

It is useful to move northward from Lake Powell along the Waterpocket Fold, comparing and contrasting the record of deformation bands from subarea to subarea along the way. The starting point is the Brimhall Bridge subarea (47 in Fig. 118A). This is physiographically and structurally a spectacular location (Fig. 122). The trail down from Hall's Overlook to Brimhall Bridge descends 300 m (1,000 ft) through Cretaceous and Jurassic strata that dip very gently east. At the base of the cliffs is a broad strike valley, along the western margin of which are discontinuous exposures of the Entrada Formation and then continuous impressive exposures of Navajo Sandstone, where it forms the limb of the Waterpocket Fold (Fig. 122). Bedding strikes N10°W and dips 40°E.

Deformation bands in the Entrada Formation are noncataclastic in nature, and most are of the DB2c subvariety. Many of the deformation bands are several centimeters in thickness and have trace lengths of meters, but many are simply several millimeters thick and have short, discontinuous trace lengths. Offsets are very small, and both normal and reverse offsets were observed. Ladder structures were seen in several outcrops, but they are rare, as are slickenlines. Deformation bands in the Navajo Sandstone appear to be cataclastic (DB1a) and tend to be very thin (several millimeters). Slickenlines are rare. The deformation bands in the Navajo Sandstone have millimeter- and centimeter-scale normal offset. Stereographic projections of the measured orientations of deformation bands in the Entrada Formation and the Navajo Sandstone in the Brimhall Bridge subarea (47 in Fig. 118B) reveal some systematics. The Navajo Sandstone is distinguished by conjugate-normal deformation bands that strike north-northwest, parallel to the trend of the Waterpocket Fold. Sets dip moderately to steeply southwestward, or gently to steeply northeastward. There is also a system of very steeply dipping deformation bands that strikes east-northeast, essentially perpendicular to the trend of the Waterpocket Fold (47 in Fig. 118B). Deformation bands in the Entrada Formation in the Brimhall Bridge subarea (47 in Fig. 118B) are very similarly oriented and may have the same tectonic significance as those in the Navajo Sandstone at this location.

Both the Entrada Formation and Dakota Sandstone crop out within The Post subarea (48 in Fig. 118A). The character and orientations of deformation bands are quite different in the two formations, even though they occupy the same structural position with respect to the Waterpocket Fold. Bedding strikes N50°W and dips 39°NE. Deformation bands in the Entrada Formation are wavy and anastomosing; they are of the DB2c noncataclastic subvariety (see Appendix Table 1). They reveal both normal and reverse separations, but no slickenlines. Deformation bands in Dakota Sandstone are white to silvery or frosted in appearance, very well organized, and cataclastic in nature. They have associated ladder structure and slickenlines and are as thick as 30 cm (12 in.). The stereographic pattern of the measured orientations of deformation bands in the Entrada Formation in The Post subarea closely resembles that for the Entrada Formation to the west, along the southeastern margin of the Kaiparowits basin (compare, for example, the Entrada deformation bands in subarea 48 [Fig. 118B] with those of subareas 36, 37, and 39 [Fig. 108B]). In specific, the stereographic projection of deformation bands in the Entrada Formation in The Post subarea reveals one main set, striking west-northwest and dipping moderately to the southwest (48 in Fig. 118B).

The Surprise Canyon subarea (49 in Fig. 118A) primarily contains uppermost Navajo Sandstone. Bedding strikes N40°W and dips 36°NE. Some of the largest deformation bands in the subarea are parallel to bedding. They are cataclastic and of the DB1d subvariety, related to flexural slip on the monocline. Most of the remainder of the deformation bands, although abundant, are tiny (just several millimeters thick) with mostly normal offsets. They appear to be noncataclastic and of the DB2a subvariety. The few reverse offsets might well be the result of rotation of early normal-slip deformation bands during monoclinal folding. With the exception of several deformation-band surfaces that appear to be strike-slip tear adjustments, there are no slickenlines or slickensided surfaces. Overall, the stereographic display of measurements of deformation bands in the Surprise Canyon sub-

Figure 122. North-directed view of the Waterpocket Fold from Hall's Overlook. White resistant sandstone is Navajo Sandstone. Lower, darker hills to right (east) are supported by Entrada Sandstone.

area (49 in Fig. 118B) is typical of that of the DB1a conjugate-normal subvariety, i.e., a "ball of string."

Burr Trail historically was one of the few openings in the "wall" of Navajo Sandstone along the full length of what is now Capitol Reef National Park, the other obvious opening being Capitol Gorge. Butch Cassidy, the Sundance Kid, and the Hole-in-the-Rock Gang knew the Burr Trail escape hatch and used it to their advantage when being pursued by the law. A switchback jeep road now provides an unusually steep access road to the bedrock surface of the monocline. Burr Trail subarea (50 in Fig. 118A) is a very challenging locality to interpret, in part because of the oblique exposures and in part because many of the deformation bands and zones of deformation bands appear to strike east and northeast rather than parallel to the monocline. In fact, the presence of this east-trending fabric may be the control for the "cut" through the Waterpocket Fold. The strong grain imparted by the deformation-band shear zones may have influenced erosion across the great dipping wall of Navajo Sandstone. There is no question that most of the zones of deformation bands that are present along Burr Trail are of tectonic origin, for the fabrics are cataclastic, and ladder structure, slickenlines, and slickensided surfaces are quite common. Some of the deformation-band shear zones are parallel to bedding and cross-bedding (DB1d subvariety), the result of slip and shear along bedding lamination during flexural-slip folding. Many of the zones of deformation bands, including ladder structures, in the Burr Trail subarea disclose low-dipping attitudes (Fig. 123), and in some cases the cross-bedding caught within Riedel deformation-band systems has become kinked. There is clear evidence for south-dipping zones of deformation bands that show thrust and reverse separation and, in the same outcrops, southeast-dipping zones of deformation bands that show low-angle normal offsets. Bedding in the Navajo Sandstone along Burr Trail strikes N40°W, which is nearly 20° farther toward the northwest than the overall trend of the Waterpocket Fold.

Just north of Burr Trail is the Peek-A-Boo Canyon subarea (51 in Fig. 118A), within which the Navajo Sandstone strikes N10°W and dips 55°E. The slot canyon within this subarea permits some westward access into the monocline in the Navajo Sandstone. At the very top of the Navajo Sandstone there are noncataclastic DB1a conjugate-normal deformation bands. Stratigraphically deeper there are outcrops containing long and continuous, rather thick, zones of deformation bands of a cataclastic nature. Some of these exhibit slickenlines and slickensided surfaces as well as some ladder geometries. I have the impression that the outcrops record both normal-slip and thrust-slip zones of deformation bands and that the thrust bands are lower on the folded arc of the monocline. Stereographic projection of the measured deformation bands make it clear that the system of cataclastic deformation bands strikes north-northwest (51 in Fig. 118B), parallel to the trend of the

Figure 123. Low-angle deformation bands cutting Navajo Sandstone along the Burr Trail. Geologists are Angela Smith and Kerry Caruthers.

Waterpocket Fold. The exact same pattern shows up just to the north in the Lizard Canyon subarea (52 in Fig. 118), even though the bedrock within which the deformation bands occur is Entrada Formation.

Farther to the north along the Waterpocket Fold is the Dogwater subarea (53 in Fig. 118A). There we find that the bedrock is gently dipping Entrada Formation. The Entrada Formation contains abundant noncataclastic DB2c deformation bands that are physically identical to those in Entrada Formation at Dance Hall Rock. The deformation bands are braided and anastomosing, and they tend to be relatively thick, up to 15 cm. Slickenlines were seen only on one surface. Offset was discernible for only one deformation band and proved to be normal slip. The stereographic pattern of measured deformation bands at the Dogwater subarea (53 Fig. 118B) resembles patterns in Entrada Formation along the southeastern margin of the Kaiparowits basin: namely, east-trending conjugate-normal deformation bands (36 and 39 in Fig. 108B).

The Sheets Gulch subarea

The Sheets Gulch subarea (54 in Fig. 118A) is one of the most unusual and impressive in the entire region of study. The Navajo Sandstone is the host rock for what may be one of the world's best expressions of Riedel strike-slip geometries, here in the form of *conjugate* zones of strike-slip deformation bands (Davis et al., 1999). The Sheets Gulch subarea occurs at the junction between the Circle Cliffs uplift on the south and the Miners Mountain uplift on the north, where the Waterpocket Fold (as traced to the north) makes a 1 km (0.6 mi) jog to the east (Figs. 46, 47, and 118A). In this salient, Navajo Sandstone dips ~15°S, and even from a distance, deformation-band shear zones are evident (Fig. 124). To the north of Sheets Gulch, the Navajo Sandstone strikes ~N20°W and dips ~25°NE, whereas to the south of the Sheets Gulch subarea, the Navajo Sandstone strikes ~N20°W and dips 55°NE. The Sheets Gulch subarea is an expansive display of slickrock Navajo Sandstone (Fig. 124) within which the details of the system of deformation-band shear zones can be mapped.

There are hundreds, probably thousands, of zones of cataclastic deformation bands in the Sheets Gulch subarea, ranging in size from those centimeters in trace length and several millimeters thick to those that are hundreds of meters in trace length and many meters thick. Photomicrographs of deformation bands in the Sheets Gulch area confirm that the fabrics are cataclastic in nature. Not only is grain fracturing abundant within thin sections, cataclastic flow is commonplace.

The largest deformation-band shear zones in the Sheets Gulch area stand out in aerial photographs (Fig. 125) and virtual images (Fig. 126) of the Sheets Gulch subarea. Long, thick, continuous zones of deformation bands project upward as fault fins or as dike-like features from the slickrock landscape, and they help resist erosion of the flanks of buttes and towers (Fig. 127). These resistant bands have a significant influence on local geomorphology, such as controlling the locations and orientations of straight stretches of canyon rims and in places causing stream piracy. A striking example of stream piracy is revealed in the northeastern part of the Sheets Gulch subarea, where a swale-like former drainage channel is perched at a relatively high elevation (Fig. 128). This former channel is incised transversely by a steep-sided slot-canyon tributary to Sheets Gulch, whose orientation and position are controlled by an especially thick deformation-band shear zone. Downcutting by the former channel was impeded by the resistance of the deformation-band shear zone it had to cross. Downcutting by the present tributary to Sheets Gulch through the host Navajo Sandstone was unimpeded and relatively rapid. The now spatially disconnected segments of the former channel have the same transverse and longitudinal profile characteristics, confirming the match (Fig. 127).

Resistance of the zones of deformation bands to erosion is also expressed as "block fields" of angular manganese- and iron-

Figure 124. A: West-directed view of Navajo Sandstone (south-dipping) in the Sheets Gulch area. We call the prominent rock form in the middle distance "Moby Dick." B: Closer view, showing the head of Moby Dick on the left, and resistant, dark, iron-stained deformation-band shear zones in Navajo Sandstone on the right. Arrows show locations of shear zones.

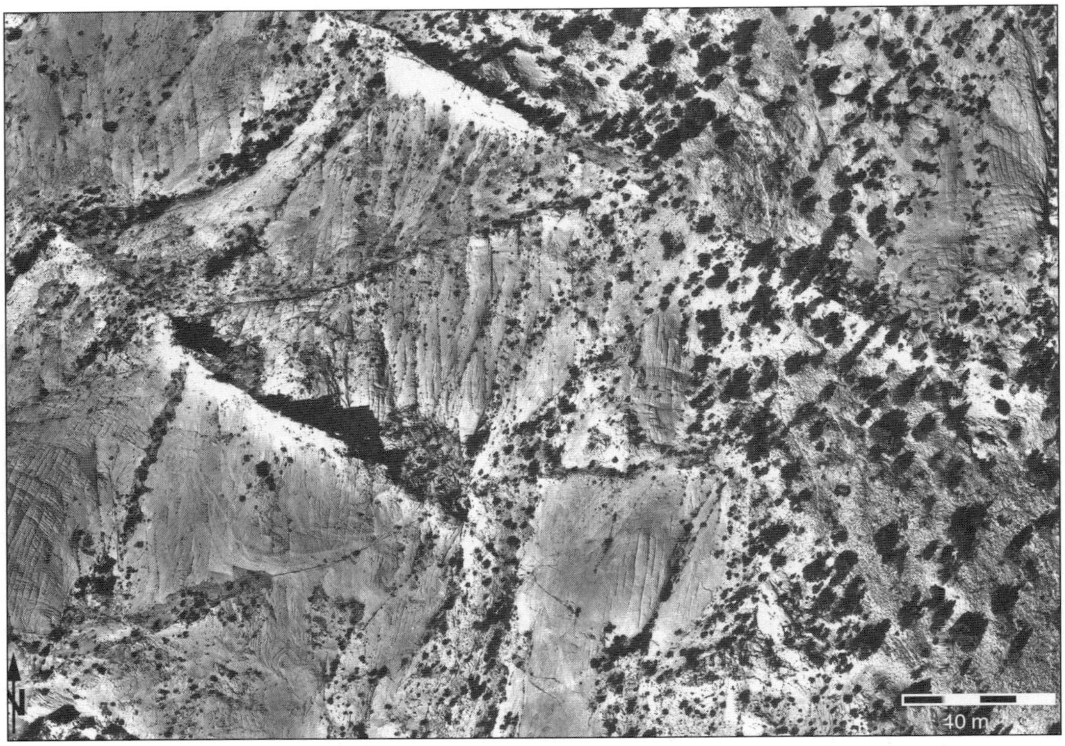

Figure 125. Aerial photographic view of small part of the Sheets Gulch subarea. Northwest-striking deformation-band shear zones are quite evident because of their resistance to erosion. Northeast-striking deformation-band shear zones are visible as well, but in this example the traces are thinner. The north-south grain is the expression of jointing, which postdates deformation banding. Individual trees are junipers, no more than 5 m (16.5 ft) in diameter.

Figure 126. Northwest-directed "virtual" oblique view of the Sheets Gulch area. Note expression of deformation bands. The image was created by Chevron through draping of our large-scale photographic imagery over our digital topography.

Figure 127. Photograph of "tower" of Navajo Sandstone in the Sheets Gulch subarea. This giant pillar of rock is flanked on both sides (right and left, note arrows) by resistant deformation-band shear zones, the upper reaches of which project as fault fins.

stained float blocks and rubble of deformation bands (Fig. 129). Individual blocks are almost wholly composed of massive deformation-band shear zones, with small ellipsoidal to lensoidal vestiges of Navajo Sandstone that somehow survived cataclastic annihilation (Fig. 130). Such block fields are found at the base of steep slickrock ravines that empty onto relatively flat slickrock terraces. Sand is moved out of the system swiftly and easily, leaving behind the resistant blocks of deformation bands.

It was for purposes of detailed structural analysis of the Sheets Gulch system of deformation bands that I subcontracted an aerial photographic flyover of a nearly 1 km^2 (0.4 mi^2) area, including preflight field placement of GPS-surveyed targets. As a result, 1:600 aerial photographs with surveyed control points were used as a basis of generating 1:600 topographic map coverage (10 ft [3.3 m] contour interval). Given the fact that the area contains nearly 250 m (825 ft) of topographic relief, the topographic base control is extraordinarily effective for the careful mapping of the deformation-band shear zones and for a reasonably comprehensive assessment of their three-dimensional characteristics. Figure 131 presents an example of the mapping carried out by my students and me.

Structural geologic mapping in the Sheets Gulch subarea resulted in the recognition of a system of conjugate strike–slip cataclastic deformation-band shear zones (Davis, 1996b; Davis et al., 1997, 1999). Two sets intersect at ~60° (Fig. 131). One set, which is right lateral, strikes ~N45°–65°E and dips ~65°–85°NW. The other set, which is left lateral, strikes ~N60°–85°W and dips ~60°–85°NE. The strike-slip nature of these zones of deformation bands is quite obvious on the ground, for bedding and cross-bedding are commonly offset in right-handed or left-handed fashion along deformation bands whose outer surfaces are commonly slickensided and marked by low-raking slickenlines (Fig. 132). Many of the slickensided surfaces are elegant in polish and presence of grooves and slickenlines. Also, there are systematic left-handed and right-handed offsets of deformation bands, the expressions of a progressive conjugate faulting. Offsets range from millimeters to meters, and there appears to be a direct relationship between deformation-zone thickness and offset: offset is ~2× to 8× zone thickness.

The most appealing dimension of the Sheets Gulch system of deformation bands is the elegance of the Riedel geometries (Davis et al., 1997, 1999). Individual principal shear zones (PSZs) are seen to be composed of a family of Riedel (R) synthetic zones of deformation bands that are aligned ~15° from the overall trend of the PSZ, in the manner shown in Figure 133. Indeed, the overall PSZs themselves, which can be traced easily and clearly by virtue of their resistance to erosion across the slickrock surfaces, have an outcrop width and a trend that is controlled by the tip points of the overstepping R surfaces (Fig. 133). The R shears are connected by transfer zones within which are abundant antithetic R′ surfaces. The transfer zones take the form of "ladder structure" geometry. The R zones are synthetic and the R′ zones are antithetic with respect to offset along the overall PSZ (Fig. 133).

Riedel relationships, and the presence of transfer zones, are impressive in plan-view slickrock exposures, both at the map scale (Fig. 131) and at the outcrop scale (Fig. 134). The careful tracing of 10s of outcrop photographs taken from waist level resulted in controlled collage cartoons, such as that shown in Figure 135, revealing the significant continuity of the Riedel shear systems and the close spacing in plan view of the transfer zones. In contrast, photographs of the strike-slip deformation-band shear zones as exposed in vertical cliff faces reveal patterns of long, continuous, massive deformation bands with relatively widely spaced transfer zones (Fig. 136).

The systematic orientations of the deformation-band shear zones in the Sheets Gulch subarea are well displayed stereographically (subarea 54 in Fig. 118B). Two dominant sets are apparent, both of which are steeply north dipping. One set strikes east-northeast, and the other strikes west-northwest. Each set is accompanied by nearly horizontal slickenlines (54 in Fig. 118B). The great circles represent both R and R′ shears. The tight organization reveals that R shears depart by ~15° from the mapped traces of the macroscopic deformation-band shear zones, within which the Riedel relationships occur (Fig. 131).

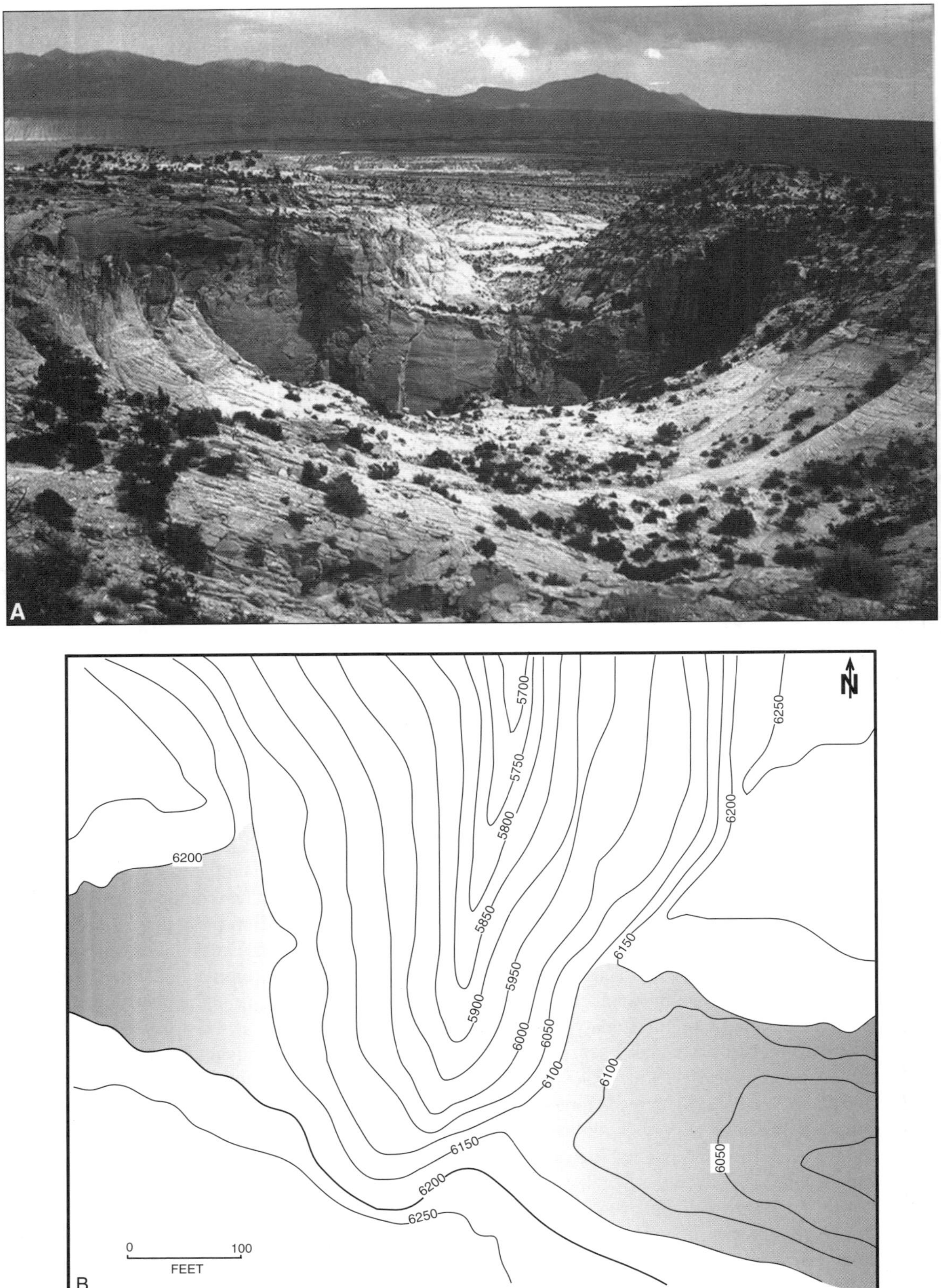

Figure 128. A: Eastward view of perched drainage transected by a deep slot canyon. The original continuity of the now-perched drainage is evident. The trace of the "pirate" canyon is in the middle ground, running from right to left. B: Topographic expression of this relationship. Shading marks old drainage. Contour interval equals 50 ft (16.5 m).

Figure 129. North-directed photograph of field of float blocks of deformation-band shear zones. The source is deformation-band shear zones farther up the canyon. The blocks are much more resistant than the undeformed Navajo Sandstone host rock in the source area. Blocks in foreground are ~0.5 m × 0.5 m.

Figure 130. Close-up view of one of the blocks in the area shown in Figure 128. Light-colored rock is cataclastically deformed Navajo Sandstone, pervaded by deformation banding. Lenses of dark (oxide-stained) rock are vestiges of largely undeformed Navajo.

The strike-slip deformation-band shear zones in the Sheets Gulch subarea actually appears to have developed before tilting of the Navajo Sandstone. When the great circles and slickenlines of the Sheets Gulch subarea (54 in Fig. 118B) are rotated by an amount consistent with removing the 15°S bedding dip along this segment of the Waterpocket Fold, the deformation-band shear zones tend to become vertical and the slickenlines tend to become horizontal (Davis et al., 1999). Strike-slip deformation-band shear thus predated the monoclinal flexing or accompanied the very early stages of folding. This conclusion is further confirmed when noting that the trace of slickenlines is always parallel to the trace of bedding (not cross-bedding) on the fault surface, which indicates that at the time of faulting, the Navajo Sandstone was horizontal and the fault movements were horizontal. No amount of tilting could thus alter the parallelism.

The system of deformation-band shear zones in the Sheets Gulch subarea has profoundly influenced fluid flow, as seen in the intense manganese and iron staining at the margins of porous, permeable host rock and nonporous, impermeable deformation bands. The staining is evident in outcrops (e.g., Fig. 134A). Thin sections reveal that porosity is essentially lost within the cataclastically reduced deformation-band shear zones. In some cases, it is evident that fluids moved through the chambers of transfer zones, sometimes succeeding in clogging up the once-porous host with secondary precipitates.

Properties of deformation bands in the northern part of the Waterpocket subregion

The Fivemile Wash subarea (55 in Fig. 118A) is located ~3 km (2 mi) north of the Sheets Gulch subarea. Bedrock is subhorizontal Entrada Formation, and the zones of deformation bands within the Entrada Formation physically resemble the noncataclastic DB2c subvariety at Dance Hall Rock. The deformation bands are very abundant. Most are very thin (several millimeters), but there are many that are 1 or 2 cm (0.5 or 1 in.) in thickness. The thickest zone of deformation bands observed in this subarea was 8 cm (3 in.). Slickenlines are rare, but where seen they rake steeply. Offsets are normal. Stereographic projection of the measured orientations reveal that the dominant system of deformation bands at the Fivemile Wash subarea (55 in Fig. 118B) strikes east and dips steeply both to the north and to the south. Not as rigorously preferred in orientation are deformation bands that strike north to northeast, dipping both northwest and southeast.

At the Notom subarea (56 in Fig. 118A), the bedrock is Entrada Formation, which here lithologically is a white sandstone that is very soft in outcrop, yet forms a bench in the landscape. Deformation bands are well developed, and most range from 1 mm to several centimeters (0.1 to 1 in.) thickness. The thickest zone observed was 30 cm (12 in.). Offsets are normal, though no slickenlines were seen. Stereographic projection of measured orientations of deformation bands at the Notom subarea (56 in Fig. 118B) reveals a system of deformation bands striking ~N35°–55°E and dipping at ~0°–70°, both to the southeast and to the northwest.

At the Paradise Creek subarea (57 in Fig. 118A), the uppermost part of the Navajo Sandstone is exposed and contains very thin, noncataclastic (DB2) deformation bands (several millimeters thick, on average) with hardly any discernible offset. Just a couple of normal offsets were observed. The stereographic patterns are typical of the DB2a subvariety of deformation bands; dominant sets strike north-northwest and northeast (57 in Fig. 118B).

My students and I spent a lot of time exploring for deformation-band shear zones along the northern part of the "Waterpocket Fold" in the stretch between Sheets Gulch and the Visitors Center for Capitol Reef National Park (Fig. 118A), but with negligible results. For example, canyon exposures of the Navajo Sandstone

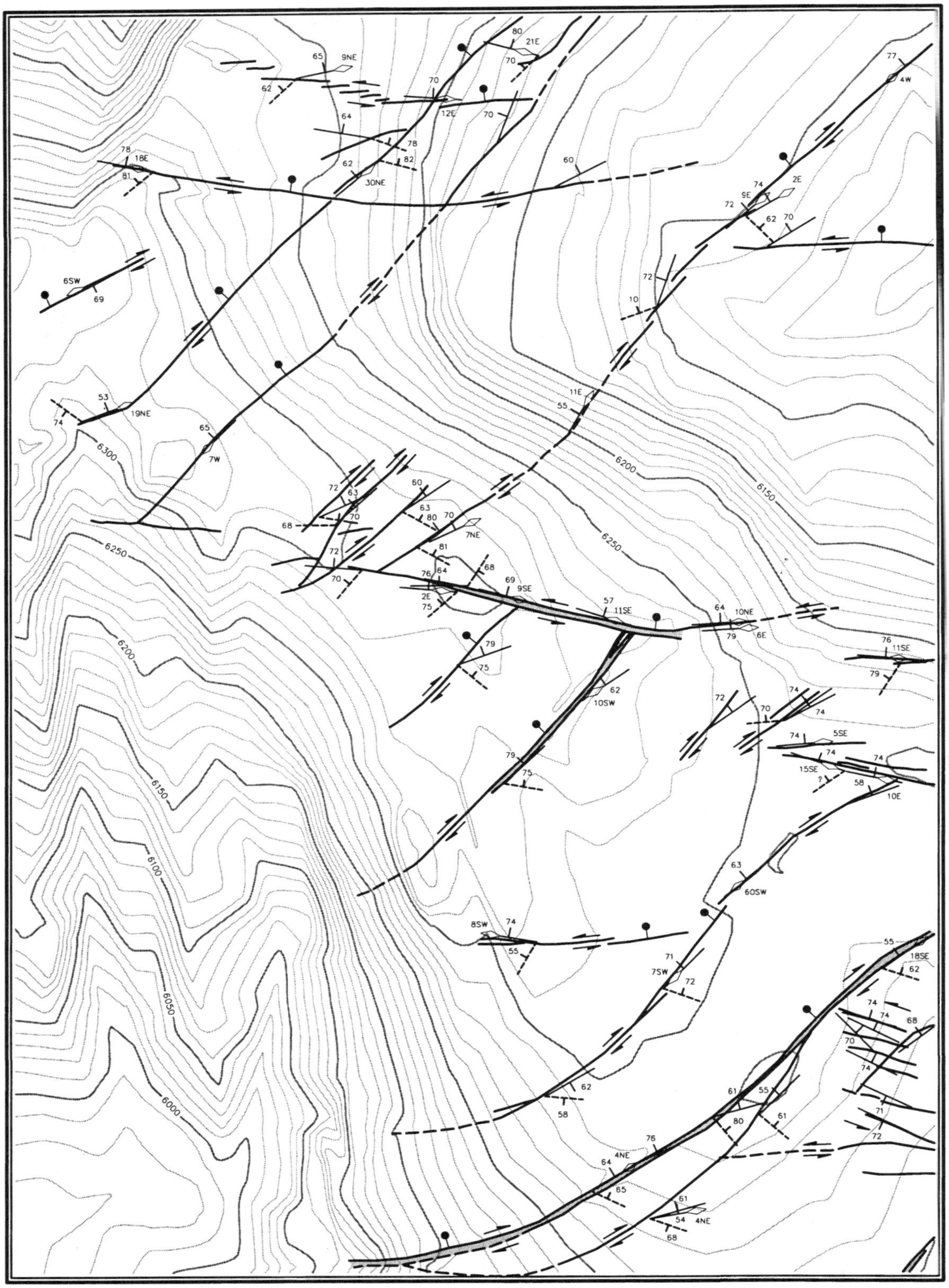

Figure 131 (this and facing page). Sample of the structural geologic mapping carried out in the Sheets Gulch subarea.

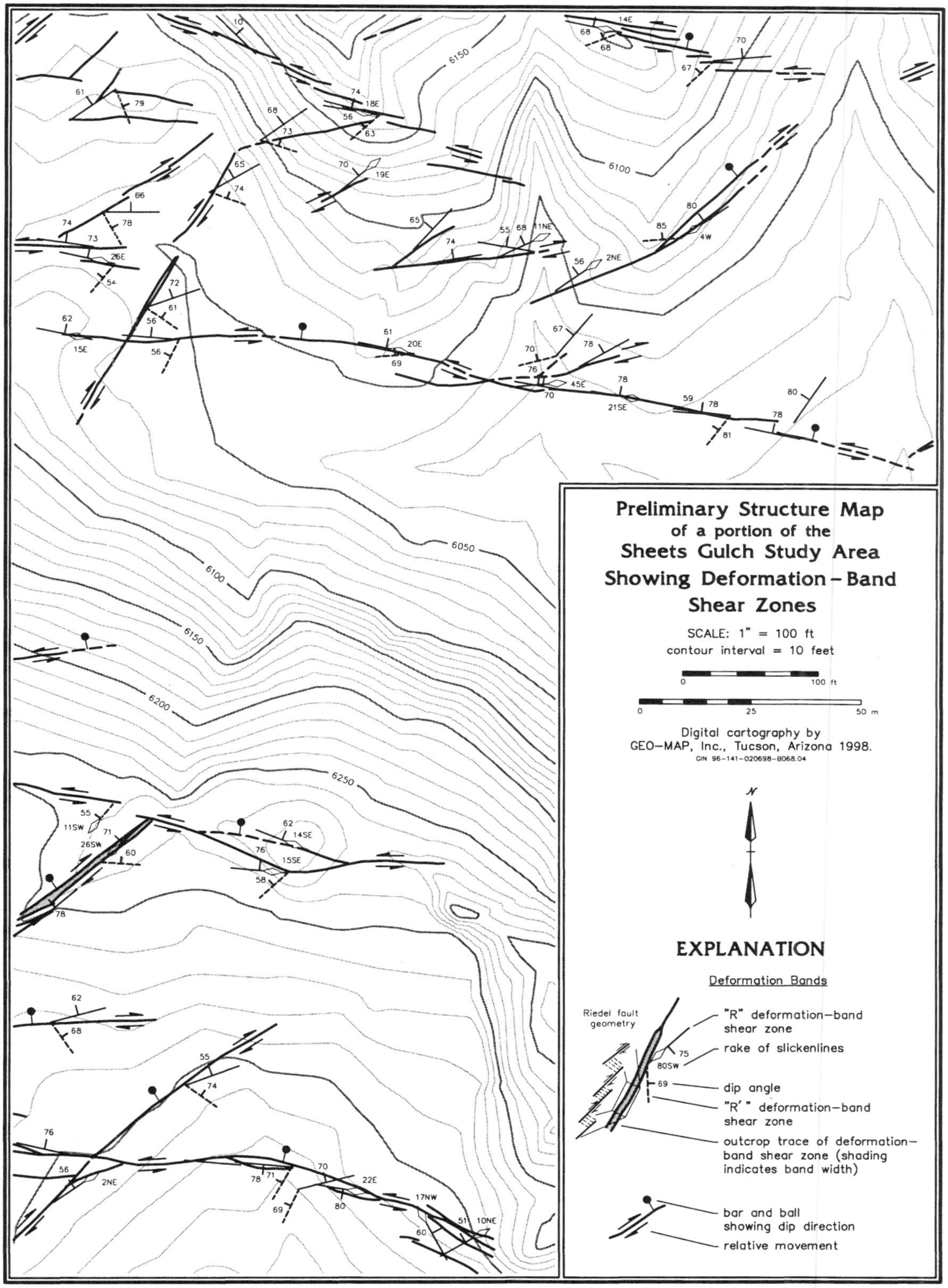

Preliminary Structure Map
of a portion of the
Sheets Gulch Study Area
Showing Deformation – Band
Shear Zones
SCALE: 1" = 100 ft
contour interval = 10 feet
0
100 ft
0
25
50 m
Digital cartography by
GEO-MAP, Inc., Tucson, Arizona 1998.
GIN 96-141-020698-8068.04
N
EXPLANATION
Deformation Bands
Riedel fault geometry
"R" deformation–band shear zone
rake of slickenlines
dip angle
"R′" deformation–band shear zone
outcrop trace of deformation–band shear zone (shading indicates band width)
bar and ball showing dip direction
relative movement
6050
6100
6150
6200
6250

Figure 132. Exquisite example of subhorizontal slickenlines on slickensided outer surface of deformation-band shear zone in the Sheets Gulch subarea. Field notebook for scale.

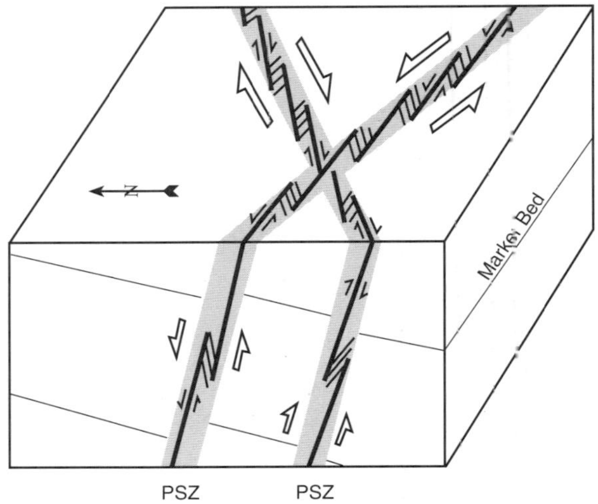

Figure 133. Line drawing showing long, continuous, massive bands in a conjugate Riedel arrangement. Transfer zones occur at locations of overstepping of individual deformation-band shear zones. Note that in this case of strike-slip faulting, the spacing between transfer zones is much greater in cross-sectional view than in plan view.

along Capitol Gorge near the Visitors Center display well-developed zones of DB1ac deformation bands at only one or two locations, which correspond with traces of faults. We thought at first that our inability to find deformation bands of tectonic origin along this part of the fold might have been related to the impossibility of gaining access to the tops of the steep cliffs in the gorge. Our serious attempts to find the source for deformation-band shear-zone float (which proved to be related to a single fault zone), for example, were thwarted by inaccessible or dangerous cliffs. Only in 1997, after the work of Bump et al. (1997), did it become clear that the dipping strata in this part of Capitol Reef National Park *do not* reflect folding along the Waterpocket. Instead, the dipping strata represent the eastward back tilt of the Miners Mountain uplift (Figs. 46 and 118A). There is negligible bending strain and thus negligible deformation-band shearing.

The Capitol Wash subarea (58 in Fig. 118A), just south of the Visitors Center, represents a complete traverse through the Navajo Sandstone where it strikes ~N10°–15°W and dips gently to moderately east-northeast. The single expression of deformation bands along Capitol Wash occurs near the Tanks (natural waterpockets), where there is a significant fault zone that strikes N62°E and dips 74°NW, with north-side-down offset. Slip on the fault is oblique, as recorded by very shallowly plunging slickenlines. Zones of deformation bands within this zone are as thick as 2 m (6.6 ft). Also near the Tanks are incredible float blocks containing deformation-band shear zones (Fig. 61C). These blocks stand out in part because they are heavily caked with iron and manganese stain and because they display boxwork architecture reflecting ladder structure. The staining once again records the degree to which deformation bands served as barriers to fluid flow, thus concentrating precipitates from the fluids that passed along the interface between the impermeable deformation-band fault rock and the porous Navajo Sandstone host rock. Outside of the zone of faulting near the Tanks, the bands are very thin (2–3 mm) and not marked by any slickenlines, nor any discernible offset. The stereographically plotted measured orientations of deformation bands in the Capitol Wash subarea (58 in Fig. 118B) are extremely well organized into two orthogonal systems, one parallel to the strike of the Navajo Sandstone (north-northwest) and the other perpendicular (east-northeast).

Origin of deformation bands in the Waterpocket subregion

The earliest deformation bands to form in rocks in the Waterpocket subregion were the noncataclastic conjugate-normal

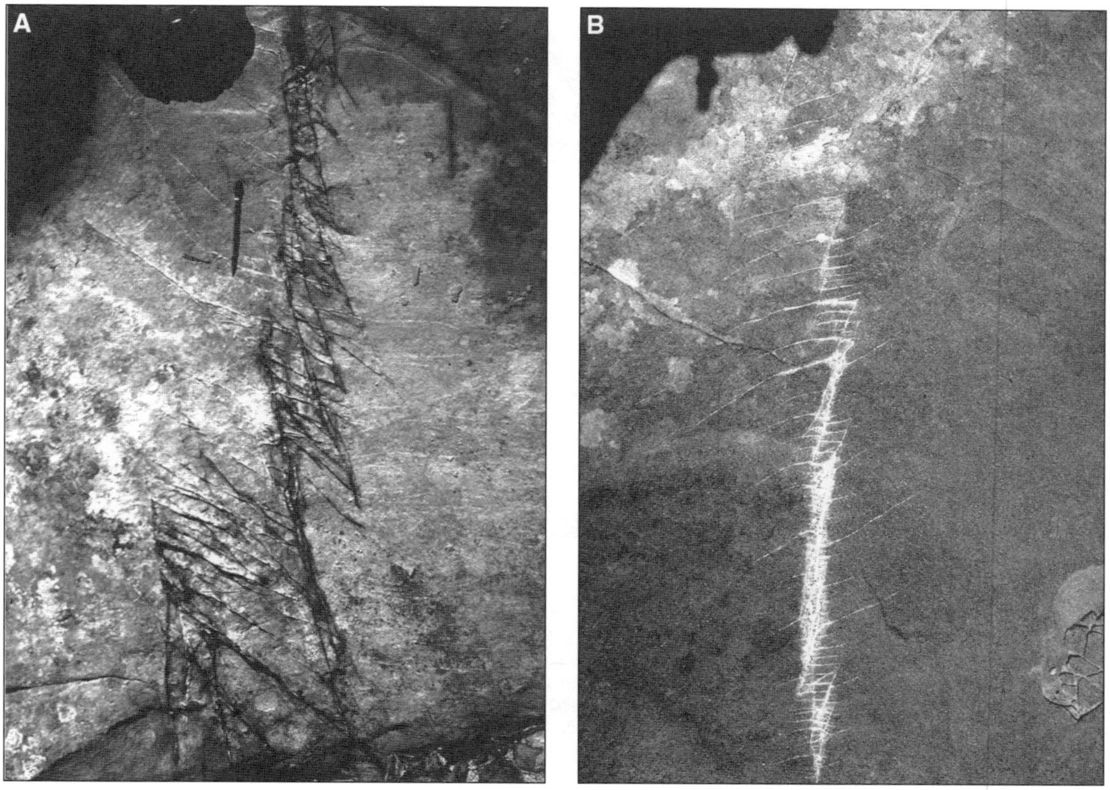

Figure 134. A: East-southeast view of outcrop-scale example of left-lateral Riedel system in the Sheets Gulch subarea. Note pencil for scale. B: Northeast view of outcrop-scale example of right-lateral Riedel system in the Sheets Gulch subarea. The frame captures a 2 m (6.6 ft) length of outcrop.

Figure 135. The Riedel pattern displayed in this cartoon is based on careful tracing of a series of photographs taken along the zone and then framing the results in an oblique perspective. Total trace length is ~10 m (33 ft).

Figure 136. Northeast-directed photograph of deformation-band shear zones (dark) exposed in near-vertical cliff of Navajo Sandstone in the Sheets Gulch area.

deformation bands of the DB2a subvariety in uppermost Navajo Sandstone, best exemplified by the patterns in the Surprise Canyon subarea (49 in Fig. 118B). Pretectonic bands of this type are interpreted to have formed through gravitational loading during burial and compaction of the Navajo Sandstone, in the ways described earlier. The greatest principal stress (σ_1) was vertical. Because the process of formation of these deformation bands involved volume loss, there was no requirement for horizontal extension. Strain was axial and vertical.

Deformation bands in the southern part of the Waterpocket subregion (Annie's Canyon, Bullfrog, Amy's, and Brimhall Bridge subareas [44 to 47 in Fig. 118A]) occur both in the Entrada Formation and in the Navajo Sandstone. On the basis of their geometric relationship to the Waterpocket Fold and related folds, these deformation bands appear to be tectonic in origin. Annie's Canyon subarea (44 in Fig. 118A) lies immediately east of the lower hinge of the Waterpocket Fold and is associated with a fault that originates in the Waterpocket Fold. The stress environment was marked by a subhorizontal, east-northeast–west-southwest orientation of the intermediate principal stress (σ_2), which was compressive and parallel to the regional direction of Laramide shortening. The greatest principal stress (σ_1) was vertical, and the least principal stress (σ_3) was subhorizontal, oriented north-northwest–south-southeast. The deformation bands at Bullfrog subarea and Amy's subarea formed under conditions of a vertical orientation of the greatest principal stress (σ_1), and the subhorizontal orientations for the intermediate (σ_2) and the least (σ_3) principal stresses, were northwest-southeast and northeast-southwest, respectively. The directions of σ_2 and σ_3 may have switched periodically, giving rise to the orthogonal systems of normal-slip deformation bands. The deformation bands in these two subareas are perpendicular and parallel to the trace of the Sock Spring syncline, a Laramide feature (Fig. 118B). Deformation bands in the Brimhall Bridge subarea (47 in Fig. 118A), with some exceptions, tend to be parallel and perpendicular in strike to the axial trace of the Waterpocket Fold, suggesting a Laramide origin for the deformation bands.

In the north-central part of the Waterpocket subregion, the Laramide origin of deformation bands is even more apparent, as shown in the data for The Post, Burr Trail, Peek-A-Boo, and Lizard Canyon subareas (48, 50, and 51 in Fig. 118A). Burr Trail is located precisely where the Waterpocket Fold, heading south, bends from ~N20°W to N55°W. This bend may be a response to strike-slip kinematics. Rotation of steeply eastward inclined bedding to horizontal, and simultaneous rotation of the deformation bands and associated slickenlines in The Post and Burr Trail subareas by the same amount, produces a conjugate–strike-slip system. The southeast-dipping deformation bands rotate into an east-northeast–striking, steeply south-southeast–dipping orientation marked by right-handed strike-slip kinematics. The south-dipping deformation bands rotate into a west-northwest–striking, nearly vertical orientation marked by left-handed strike-slip kinematics. The acute bisector trends west-southwest, nearly perpendicular to the overall orientation of the trend of the Waterpocket Fold. Thus the stress environment responsible for the formation of these deformation bands may have been marked by a horizontal greatest principal stress (σ_1) oriented east-northeast–west-southwest; a vertical intermediate principal stress (σ_2); and a horizontal least principal stress (σ_3) oriented north-northwest–south-southeast. This geometry suggests that conjugate sets of deformation bands were produced in Navajo Sandstone and Entrada Formation during the Laramide compression along the Waterpocket Fold but before significant monoclinal tilting of beds. Deformation bands at Peek-A-Boo and Lizard Canyon subareas (51 and 52 in Fig. 118B) strike parallel to the trace of the Waterpocket Fold and reflect a stress field marked by a subvertical greatest principal stress (σ_1), a subhorizontal intermediate principal stress (σ_2), oriented north-northwest–south-southeast, and a subhorizontal least principal stress (σ_3) oriented east-northeast–west-southwest.

As emphasized already, the location of the Sheets Gulch area (54 in Fig. 118A) is precisely where the Waterpocket Fold to the south is replaced by the southwest-vergent Miners Mountain monocline to the north (Bump et al., 1997) (Fig. 66 and 67). The Miners Mountain monocline is associated with an antithetic left-handed reverse fault (Anderson and Barnhard, 1986; Bump et al., 1997). Bump et al. (1997) have suggested that the conjugate–strike-slip system in the Sheets Gulch subarea is a strain accommodation to this transferring of displacement. The transfer was imperfect. Extra strain "leaked" into the Sheets Gulch area in the form of a distributed strike-slip deformation. Principal stress directions, of Laramide origin, for the Sheets Gulch system of deformation-band shear zones are as follows: the greatest principal stress (σ_1) was subhorizontal and oriented east-northeast–west-southwest, the intermediate principal stress (σ_2) was subvertical, and the least principal stress (σ_3) was subhorizontal and oriented north-northwest–south-southeast.

The northern end of the Waterpocket subregion again shows the imprint of a Laramide origin; the deformation bands in the Notom, Paradise Creek, and Capitol Wash subareas (56 to 58 in Fig. 118B) are aligned parallel and perpendicular in strike to the trace of the Waterpocket Fold. Horizontal compression oriented east-northeast–west-southwest is reflected in the line of intersection of the dominant sets of deformation bands, suggesting that the intermediate direction of principal (compressive) stress (σ_2) was subhorizontal and oriented east-northeast–west-southwest. The greatest principal stress (σ_1) was vertical. The intermediate principal stress (σ_2) was subhorizontal and oriented north-north-west–south-southeast. It would appear that the directions of intermediate (σ_2) and least principal stress (σ_3) switched periodically.

ANALYSIS OF DEFORMATION BANDS, SAN RAFAEL SWELL SUBREGION

General nature of the subregion

The San Rafael Swell subregion (Fig. 137A) is centered on the San Rafael Swell, a classic Laramide uplift (Fig. 138). Resis-

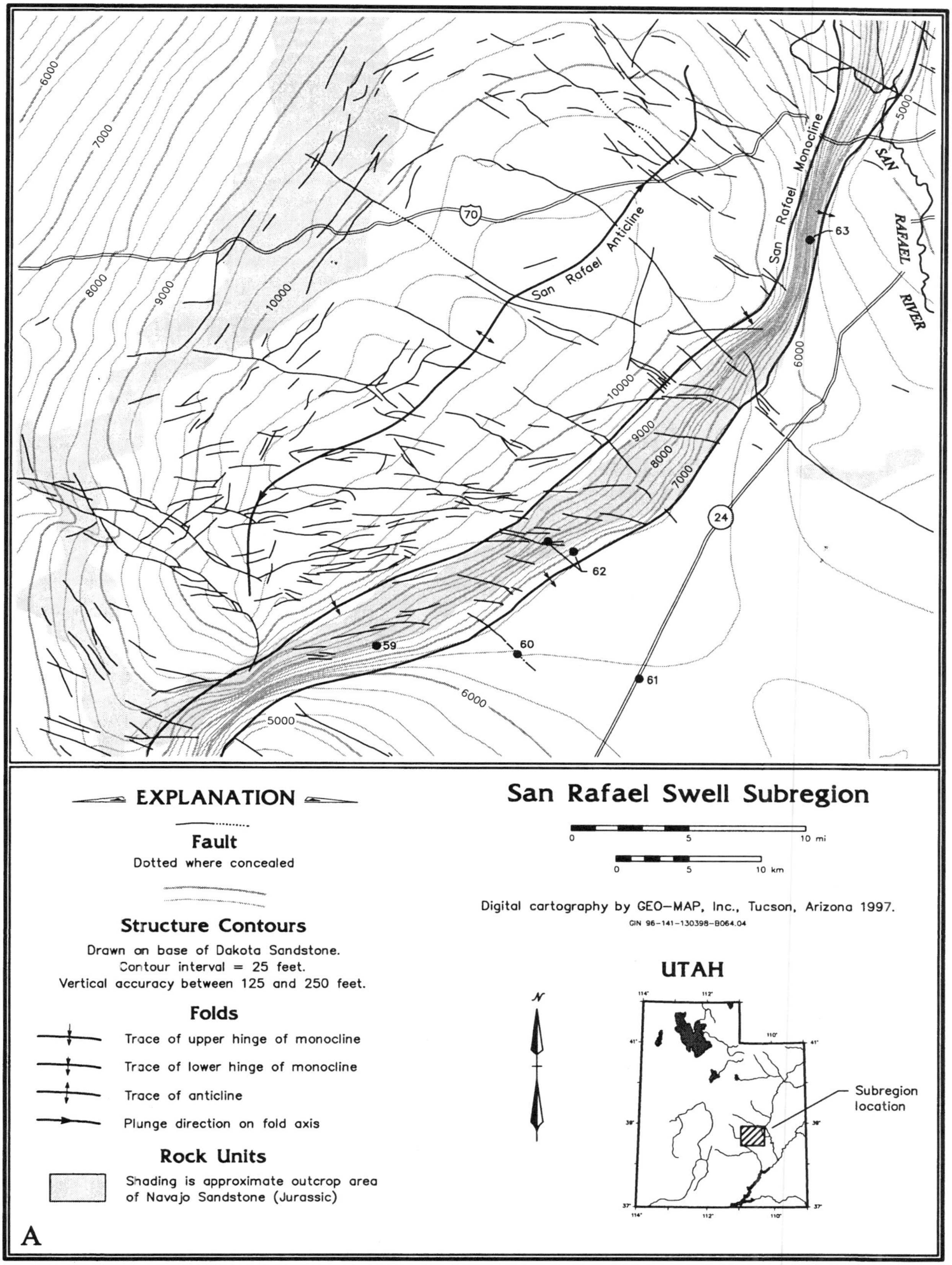

Figure 137 (this and next page). A: Map of the San Rafael Swell subregion, showing distribution of Navajo Sandstone and locations of subareas where deformation-band data were collected.

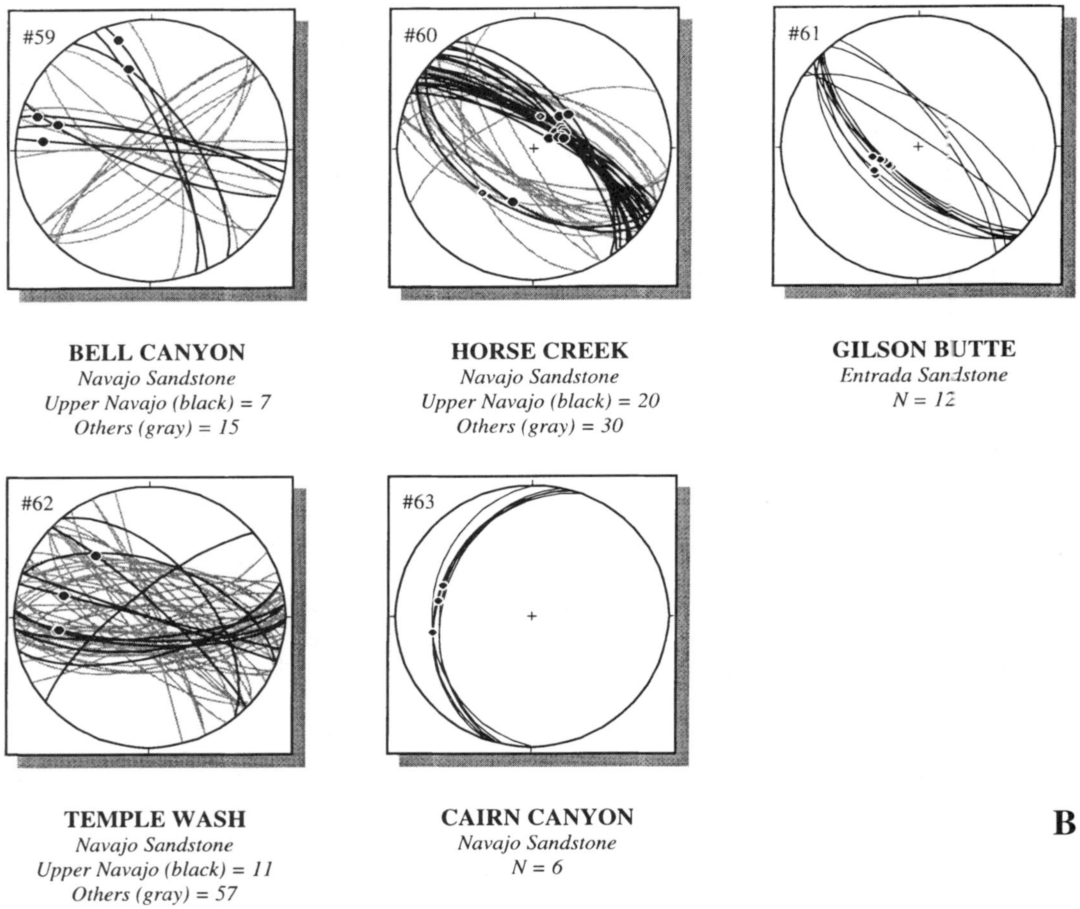

Figure 137. B: Stereographic projections of deformation bands (great circles) and slickenlines (dots) measured in the San Rafael Swell subregion. Numbers 59–63 refer to the subareas located in A. *N* = number of observations.

Figure 138. Northwest-directed aerial photographic view of the San Rafael monocline. Note the expression of the Navajo Sandstone (white) along the middle limb of the monocline. Photograph by Kurt Constenius.

tant Navajo Sandstone creates the dominant expression of the steeply dipping limb of the San Rafael monocline (Fig. 49), and the Entrada Formation is broadly accessible immediately east of the monocline, where strata are essentially horizontal (Fig. 139). The San Rafael Swell area is in many ways the type area for deformation bands, for it was here that Aydin (1978) and Aydin and Johnson (1978) carried out studies that would bring this special class of structure to the attention of the structural geology community. Deformation bands and zones of deformation bands occur both in the Navajo Sandstone and in the Entrada Formation (Appendix Table 2). My work in this region was strictly reconnaissance in nature, yet even reconnaissance studies reveal the rich record of deformation bands and zones of deformation bands. The highlights are well-exposed, closely spaced thrust faults expressed as DB1a cataclastic deformation bands and zones of deformation bands in Navajo Sandstone in the steepened limb of the San Rafael monocline; well-developed normal faults and strike-slip faults expressed as DB1a cataclastic deformation bands and zones of deformation bands in the Entrada Formation; and normal faults expressed as cataclastic deformation bands and zones of deformation bands in Navajo Sandstone on the culmination of the San Rafael Swell.

Properties of deformation bands, subarea by subarea

The Bell Canyon subarea (59 in Fig. 137A) is occupied by Navajo Sandstone. Within this subarea there are many thin deformation bands of uncertain origin, but there are also some conspicuous thick zones of deformation bands that are cataclastic and of tectonic origin. They are marked by conspicuous slickenlines, slickensides, and grooves. The slickenlines and grooves are relatively low raking, ranging from ~13° to 40°W (Fig. 137B). Bedding within the Bell Canyon subarea strikes northeast and dips southeast (see structure contours, Fig. 137A). Stereographic projection of the measured orientations of deformation bands within the Bell Canyon subarea (59 in Fig. 137B) reveals a complex system, with sets striking ~N45°E, ~N30°W, and ~N80°W. Dip inclinations are moderate to steep. These orientations are symmetrically disposed with respect to bedding.

In the Horse Creek subarea (60 in Fig. 137A), the lower part of the Entrada Formation is well exposed in desert terrain containing large patches of slickrock. This subarea lies very close to Goblin Valley State Park (Fig. 139). Some of the deformation bands in this subarea are relatively curvy and anastomosing, similar to those in the Kaiparowits subregion, with no discernible slickenlines and with few examples of discernible (normal) offset. Yet in addition to these, there are excellently exposed long, thick, continuous zones of cataclastic deformation bands that stand out in bold fin-like and wall-like topographic relief (Fig. 140). These are some of the zones of deformation bands described and interpreted by Aydin (1978) and Aydin and Johnson (1978) and more recently discussed by Fossen and Hesthammer (1997). These zones of deformation bands are as thick as 2 m (6.6 ft), and their surfaces are generally marked by well-developed slickensides and slickenlines (Fig. 140) and L-structure (Fig. 53). In any given outcrop, the

Figure 139. Photograph of Entrada Formation at Goblin Valley, east of the San Rafael Swell. Height of butte at center is ~30 m (100 ft).

Figure 140. A: Expression of deformation-band shear zone in Entrada Formation. Note dip-slip slickenlines on surface of structure. B: Close-up view of the flank of a deformation-band shear zone in the Entrada Formation. Note polish and slickenlines. Knife for scale at base of outcrop.

rake of the slickenlines and the rake of the L-structure are strictly parallel and very steep. For example, the fault surface shown in Figure 53 is exposed on the south wall of a major zone of deformation bands. The fault surface strikes N47°W, dips 73°E, and is marked by slickenlines and grooves that rake 80°W. Bedding is essentially horizontal. Offsets are normal, as recorded primarily in ladder geometries that are very well developed. Microfabrics are cataclastic in nature. The stereographic display of measured deformation bands in the Horse Creek subarea (60 in Fig. 137B) reveals strong preferred orientations. The deformation bands strike ~N40° to N60°W and dip moderately steeply both to the northeast and southwest. The deformation bands and zones of deformation bands in the Horse Creek subarea are thus parallel to the major graben-bounding dip-slip normal faults (Fossen and Hesthammer, 1997) that cross the San Rafael Swell in this general area.

The bedrock at the Gilson Butte subarea (61 in Fig. 137A) is white sandstone within the Entrada Formation, and bedding is essentially horizontal. The subarea is marked by beautiful fin-like projections of cataclastic deformation bands (DB1a subvariety) with polished surfaces, local L-structure, and slickenlines. These zones of deformation bands are as thick as 2 m (6.6 ft). Displacements are normal, and slickenlines are steeply raking. This too is a part of the area reported on by Aydin (1978) and Aydin and Johnson (1978). Orientations and kinematics of deformation bands in the Gilson Butte subarea are identical to those of the Horse Creek subarea (compare 61 with 60 in Fig. 137B): a system of northwest-striking dip-slip normal faults that are inclined both northeast and southwest.

At the Temple Wash subarea (62 in Fig. 137A), Navajo Sandstone can be sampled for deformation bands and zones of deformation bands from near the top to the bottom. Many of the bands are thin (2 mm to 2 cm, i.e., as thick as 1 in.), and especially near the bottom of the Navajo Sandstone they are highly stained with manganese. These thin deformation bands seldom are marked by slickenlines or discernible offsets. The other common variety is clearly tectonic, consisting of ladder geometries (DB1c) and slickensided, slickenlined surfaces. Kinematics are oblique slip, for the slickenlines rake ~35°W. The preferred orientation of the system of deformation bands within the Temple Wash subarea (62 in Fig. 137B) reflects the orientation(s) of the major local faults that cut through the San Rafael Swell in this general area (Fig. 137A). The deformation bands and zones of deformation bands strike west-northwest and dip moderately to steeply, both to the north-northeast and south-southwest.

One of the most significant array of zones of deformation bands in the San Rafael Swell subregion is located in a canyon I refer to as Cairn Canyon (63 in Fig. 137A). This location is in the steeply dipping limb of the San Rafael monocline. The Navajo Sandstone at this location strikes ~N25°E and dips ~55°E. A four-wheel-drive jeep trail provides access to the mouth of the canyon. The north wall of Cairn Canyon is marked by an incredible display of penetrative thrust faults, expressed in the form of zones of deformation bands spaced at distances of ~3 m (6.6 ft)

Figure 141. A: North-directed photograph of thrust-slip deformation-band shear zones in Navajo Sandstone along the San Rafael monocline in the Cairn Canyon area. Height of cliff is 50 m. B: Closer view of the thrust-slip deformation bands and the offset of marker units within the Navajo Sandstone.

(Fig. 141). I measured similar thrusts along a part of the East Kaibab monocline, but nowhere are they as well developed and well exposed as here. The deformation-band thrust zones strike ~N5°E, almost perfectly parallel to the trend of the San Rafael monocline at this location. They dip ~25°W (63 in Fig. 137B). The zones of deformation bands are marked by offsets that appear to range from 0.5 to 3 m (1.6 to 11 ft). Slickenlines and grooves rake directly down the dip of the thrust surfaces (63 in Fig. 137B). Had these particular structures been recognized a decade or more ago, they may have immediately ended any lingering debate as to whether Colorado Plateau monoclines, such as the San Rafael monocline, were the products of compressional deformation or differential vertical uplift. The compressional nature of the monoclinal folding is clearly expressed in these cataclastic deformation-band thrust zones.

A final site I studied in the San Rafael subregion is located near the Chimney Rock area, near the culmination of the San

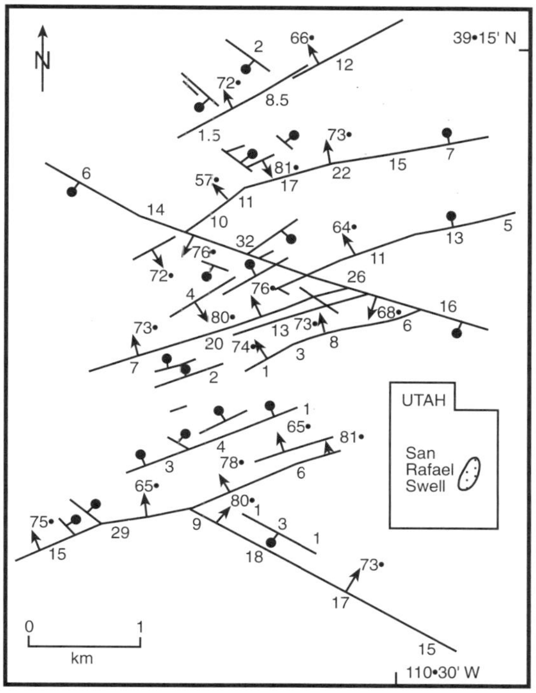

Figure 142. Structure map of faults in the Chimney Rock area of the San Rafael Swell. From Krantz (1986). The numbers, in meters, record offset from place to place along the faults. Numbers with degree symbols are dips of faults. "North fault" is the second from the top.

Figure 143. East-directed photograph of fault-line scarp of a part of North fault in the Chimney Rocks area on the San Rafael Swell. Resistant footwall is Navajo Sandstone. The height of the scarp (5 m [17 ft]) accurately reflects displacement on the fault.

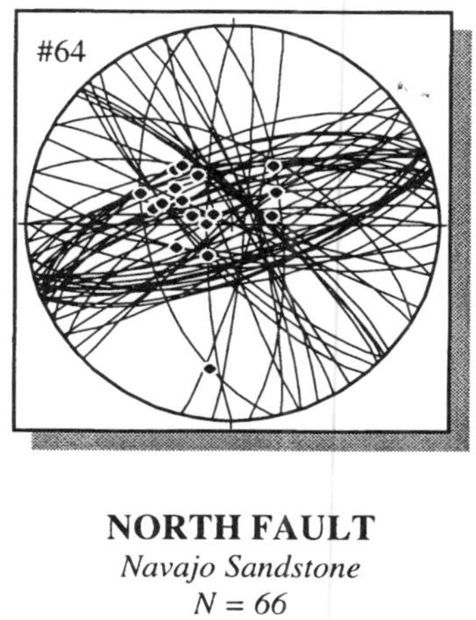

Figure 144. Stereographic projection of orientations of deformation bands (great circles) and slickenlines (dots) in Navajo Sandstone along the footwall of North fault in the Chimney Rocks area of the San Rafael Swell.

Rafael Swell (see Krantz, 1986, 1988). The area is occupied by as unusually well-exposed fault system, the basic elements of which Krantz noticed on Gilluly's (1928) geologic map of the San Rafael Swell and on aerial photographs. Krantz (1986, 1988) analyzed the fault system in great detail and concluded that the fault system conforms with Reches's (1978, 1983) descriptions of faulting in a three-dimensional strain field. The faults, although normal, have plan-view expressions that resemble conjugate–strike-slip faults (Fig. 142). The exposures of the fault system are uncanny, because fault-line scarps of footwall Navajo Sandstone accurately reflect total slip on the associated faults (Fig. 143). Individual faults were mapped by Krantz from tip point to tip point, over distances of many kilometers, and with displacements ranging from less than 1 m (3.3 ft) near the tips to more than 20 m (66 ft) near the locations of maximum slip. Moreover, the fault line scarps of Navajo Sandstone are well polished, slickenlined, and grooved, allowing slip direction to be determined *everywhere* within the fault system. Krantz (1986, 1988), through a clever and practical adaptation of Reches's analytical techniques for evaluating strain due to faulting in a three-dimensional strain field, determined that the fault system had accommodated small but measurable stretching both parallel and perpendicular to the axis of the swell.

The reconnaissance examination of deformation bands in the Chimney Rock area focused on the North fault subarea (Fig. 144). My students and I simply sampled the orientations of deformation bands and zones of deformation bands along a 300 m (990 ft) stretch of the North fault. We measured orientations of deformation-band shear zones in the immediate footwall and right along the fault surface. The fault is a normal fault that strikes ~N76°E and dips at ~70°N. The magnitude of displacement in the segment of interest is ~20 m (66 ft). The stereographically plotted data reveal that there are two systems of conjugate-normal faults (Fig. 144), one striking parallel to the trend of North fault, i.e., east-northeast, the other striking perpendicular to the main zone, i.e., northwest. Slickenlines on both systems are steeply raking.

Origin of the deformation bands

Both Laramide compressional tectonics and post-Laramide extensional faulting are recorded in deformation bands in the San Rafael Swell subregion. Evidence for strong west-northwest–east-southeast shortening, i.e., perpendicular to the trend of the San Rafael monocline, is revealed by stereographic patterns for the Bell Canyon, Temple Wash, and Cairn Canyon subareas (59, 62, and 63 in Fig. 137B). The thrust deformation-band shear zones at Cairn Canyon accommodated shortening perfectly perpendicular to the north-northeast local trend of the San Rafael monocline. The greatest principal stress (σ_1) was horizontal and oriented west-northwest–east-southeast. The intermediate principal stress (σ_2) at the time of deformation-band formation was subhorizontal and oriented north-northeast–south-southwest. The least principal stress (σ_3) was vertical. The Bell Canyon and Temple Wash subareas (59 and 62 in Fig. 136B) both disclose shortening achieved by strike-slip faulting. The greatest principal stress (σ_1) was horizontal and oriented northwest-southeast, the intermediate principal stress (σ_2) was vertical, and the least principal stress (σ_3) was horizontal and oriented northeast-southwest.

Most of the northwest-striking faults that cross the San Rafael Swell (Fig. 137A) are normal slip in nature and are thought to have been formed by post-Laramide collapse accompanying dissolution of salt of the Paradox Formation (Middle Pennsylvanian age) (Witkind, 1991, 1994). Gilluly (1928) in fact observed that the northwest-striking faults on the San Rafael Swell were the result of "vertical readjustments later than the principal folding" (see Witkind, 1991, p. 145). Dissolution took place in post-Eocene, pre-Pliocene time (Witkind, 1991). Deformation bands measured in the Horse Creek, Gilson Butte, and Temple Wash (60 to 62 in Fig. 137B) record this post-Laramide collapse. Stress directions can be reconstructed on the basis of fault orientations and slickenlines: The greatest principal stress (σ_1) was vertical, the intermediate principal stress (σ_2) was subhorizontal and oriented northwest-southeast, and the least principal stress (σ_3) was subhorizontal and oriented northeast-southwest. The deformation bands associated with North fault (subarea 64) in the Chimney Rocks area of the north-central part of the San Rafael Swell are a product of deformation within a three-dimensional strain field, in which greatest stretch was essentially north-south and intermediate stretch was essentially east-west (Krantz, 1986). This deformation was probably Laramide, given the geographic proximity of the fault system to the structural culmination of the San Rafael Swell.

ANALYSIS OF DEFORMATION BANDS, HENRY MOUNTAINS SUBREGION

General nature of the subregion

The Henry Mountains subregion contains three of the five major lobes of the Henry Mountains: Mount Ellsworth, Mount Holmes, and Mount Hillers (Fig. 145A; also see Figs. 30 and 31). Quartz diorite intruded upward through most of the sedimentary cover in early Miocene time and then spread laterally at depth to lift and dome the entire Mesozoic column as well as some of the uppermost Paleozoic strata (Jackson and Pollard, 1988a). Consequently, the centers today are marked by igneous cores surrounded by up-tilted Mesozoic formations. Structural relief as measured on the Ferron Sandstone Member of the Mancos Shale for Mount Holmes, Mount Ellsworth, and Mount Hillers is ~1,200 m (4,000 ft), 1,500 m (5,000 ft), and 1,800 m (6,000 ft), respectively (Hunt, 1988a). The Navajo Sandstone and the Entrada Formation are dominant units around the lower margins of each of these domes, where the sandstones dip moderately to steeply. Furthermore, Navajo Sandstone and Entrada Formation make up much of the flat-lying subregional strata that encompass the Henry Mountains, thus providing opportunities to sample "background" rocks for deformation-band varieties and orientations. Deformation bands and deformation-band shear zones of a cataclastic origin are especially well developed in the Navajo Sandstone where the strata have been upturned. Deformation bands of a cataclastic origin in the Entrada Formation are also especially well developed where strata have been upturned from horizontal attitudes. These deformation bands are fundamentally noncataclastic. Yet they are tectonic in nature, marked commonly by offsets and slickenlines, and more rarely by ladder structure. The microfabrics show grain fracturing but not wholesale cataclasis (see Appendix Table 2).

Properties of the deformation bands in the subregion

It is useful to describe the patterns of deformation-band shear zones subarea by subarea, starting from south of Mount Ellsworth and working to Mount Hillers (Fig. 145A). To the south of Mount Ellsworth, the Entrada Formation is very well exposed, dipping ~2°W in accord with regional dip. At the Hansen Creek subarea (65 in Fig. 145A), deformation bands in the Entrada Formation are noncataclastic (DB2) and for the most part quite thin, typically a few millimeters in thickness, although several are as thick as 1 m. Slickenlines are very rare, and where found, they rake steeply westward. Stereographic projection of attitude measurements at the Hansen Wash subarea (65 in Fig. 145B) reveals that the deformation bands strike northwest to west-northwest, and they dip moderately to steeply to the south.

Closer to Mount Ellsworth, within the Ticaboo subarea (66 in Fig. 145A), deformation bands within the Entrada Formation strike ~N45°E and dip steeply both to the northwest and to the southeast (Fig. 145B). The deformation bands appear to be noncataclastic, but offsets and slickenlines are commonly observed. Offsets are normal and slickenlines are steeply raking. The predominant ~N45°E strike of the deformation bands at the Ticaboo subarea is parallel to the set of northeast-striking faults that can be traced from Mount Ellsworth to Mount Holmes (Fig. 145A). Furthermore, the Ticaboo subarea lies directly on the line of projection of this fault zone.

Sandstone of the Entrada Formation and Navajo Sandstone

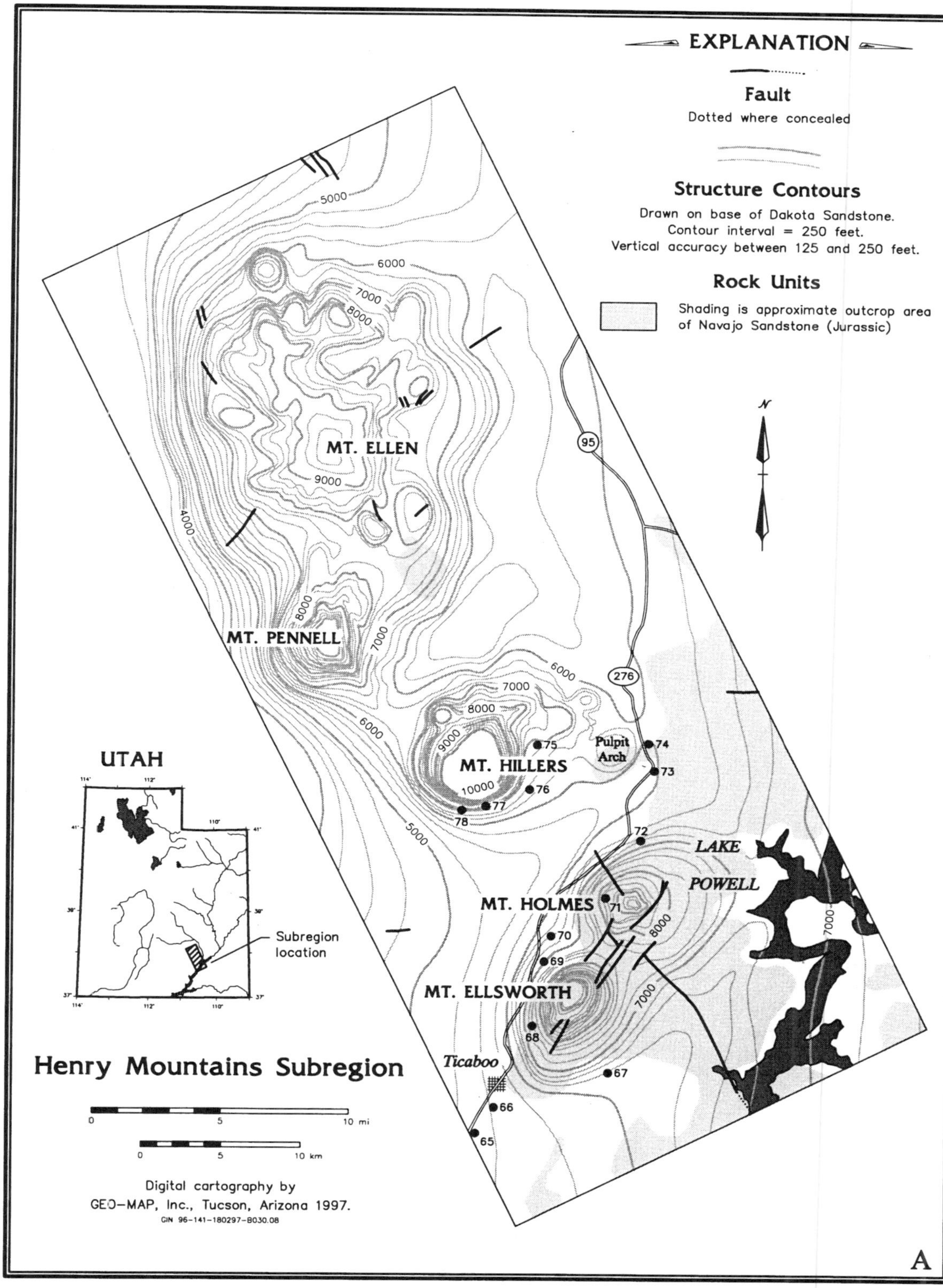

Figure 145 (this and following two pages). A: Map of the Henry Mountains subregion, showing distribution of Navajo Sandstone and locations of subareas where deformation-band data were collected.

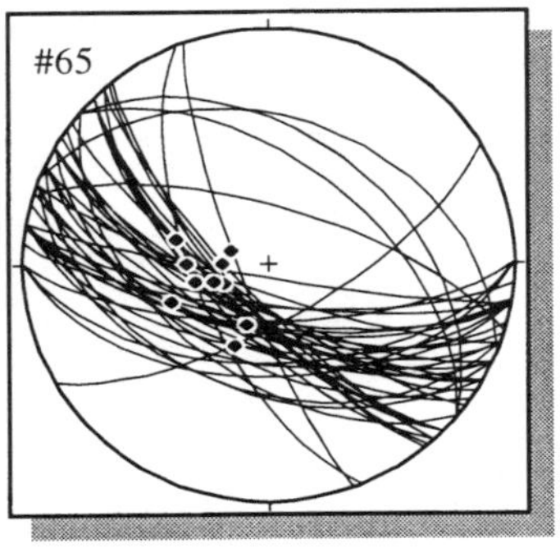

HANSEN CREEK
Entrada Sandstone
N = 47

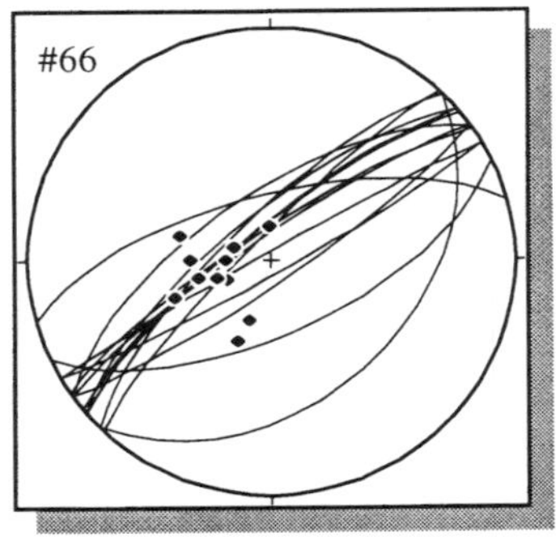

TICABOO
Entrada Sandstone
N = 14

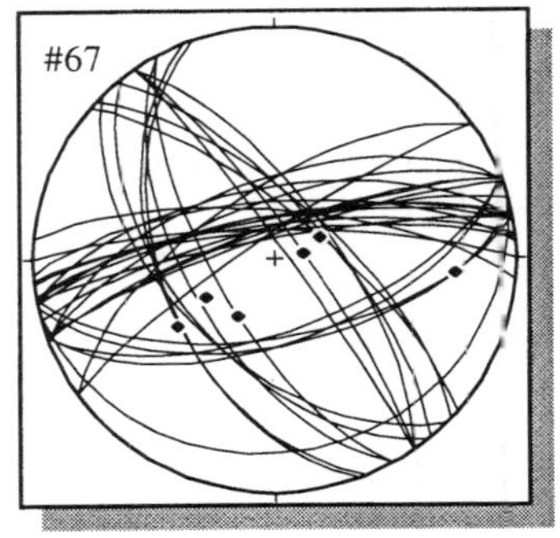

SOUTHEAST ELLSWORTH
Entrada Sandstone
N = 35

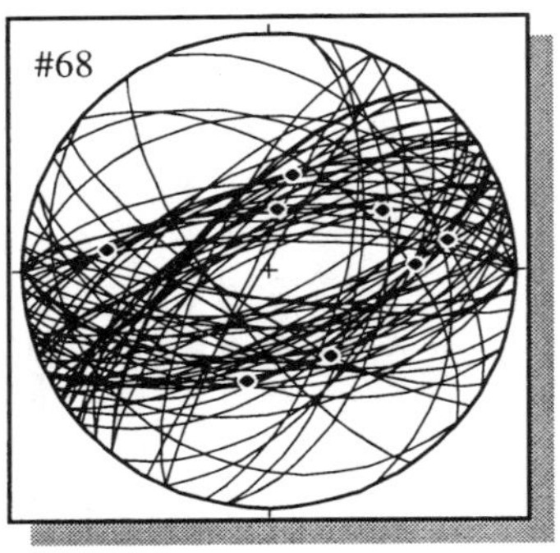

SOUTHWEST ELLSWORTH
Entrada Sandstone
N = 80

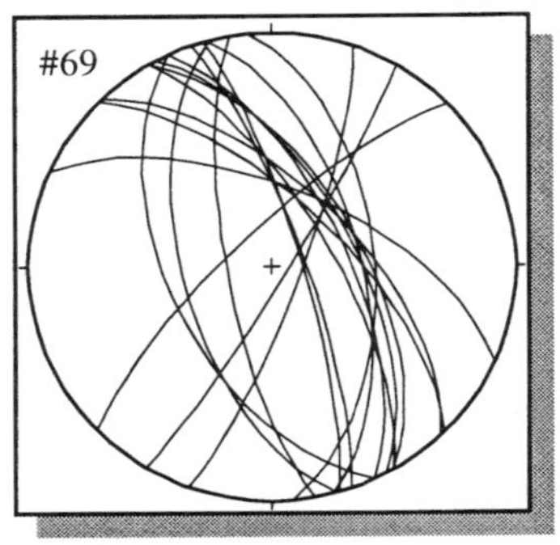

NORTHWEST ELLSWORTH
Entrada Sandstone
N = 17

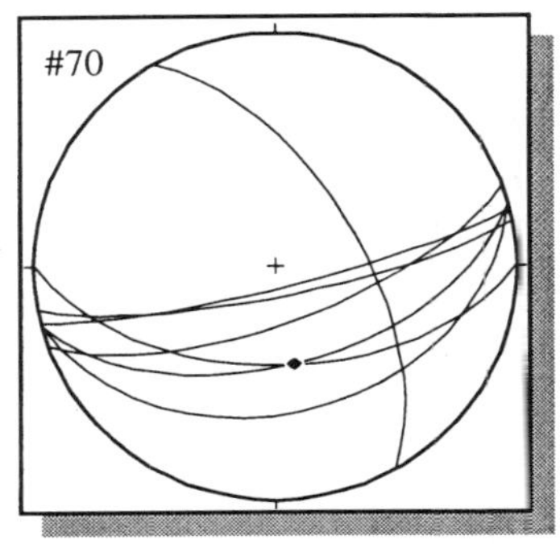

SOUTHWEST HOLMES
Entrada Sandstone
N = 7

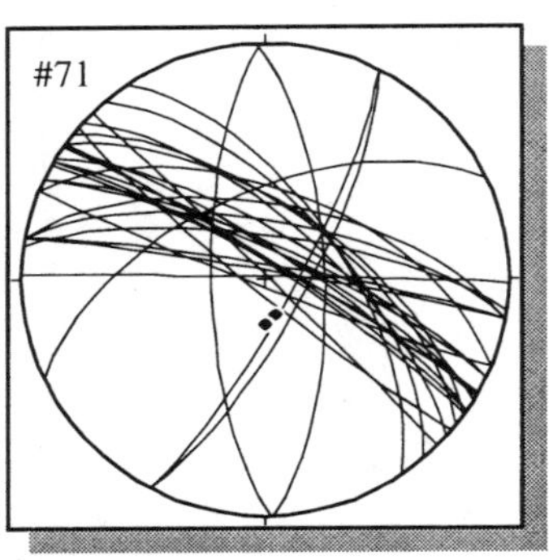

WNW HOLMES
Entrada Sandstone
N = 29

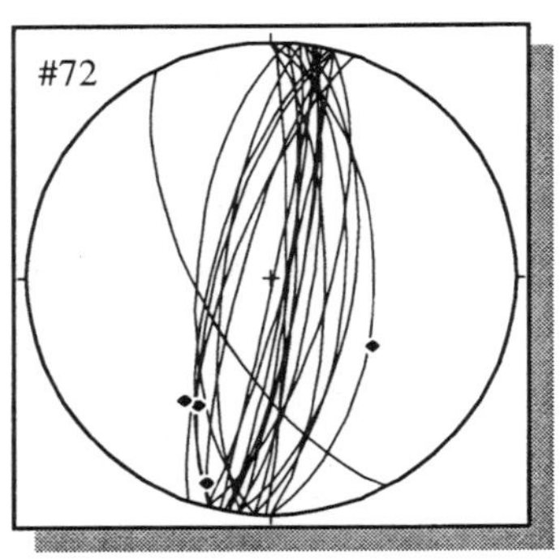

NORTH HOLMES
Navajo Sandstone
N = 21

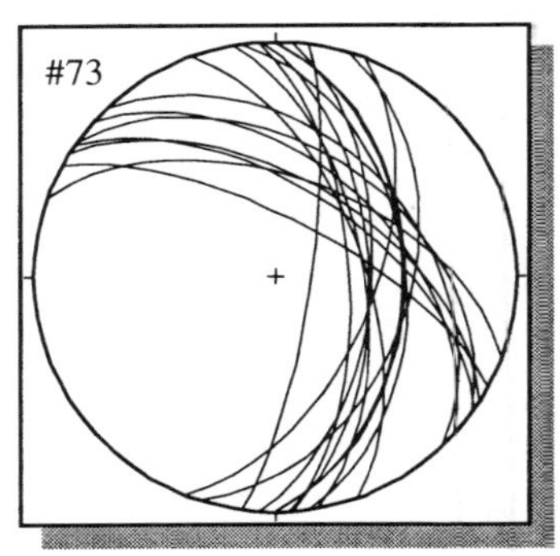

LOWER TRAIL CANYON
Entrada Sandstone
N = 19

B

Figure 145 (continued). B: Stereographic projections of deformation bands (great circles) and slickenlines (dots) measured in the Henry Mountains subregion. Numbers 65–78 refer to the subareas located in A. *N* = number of observations.

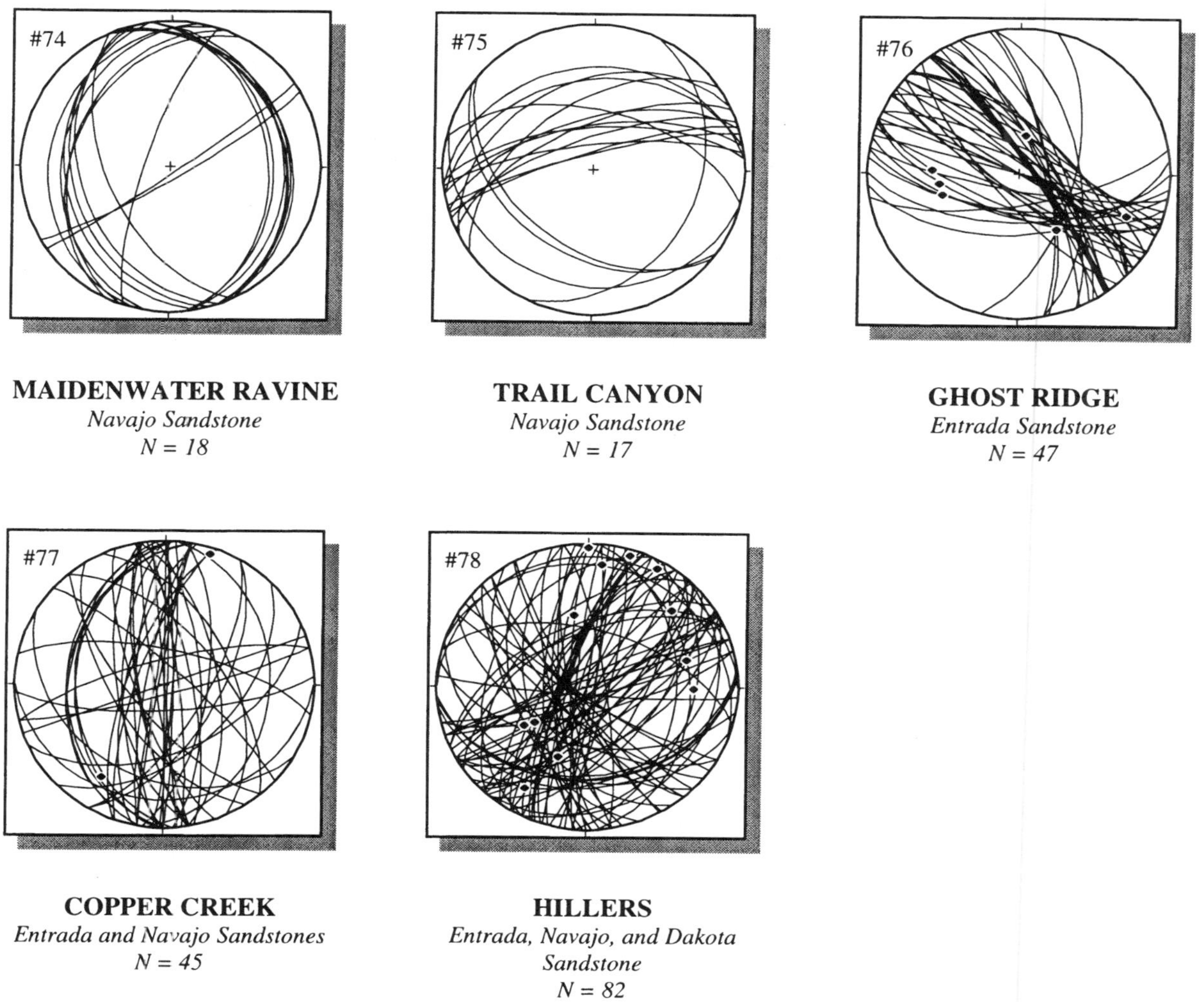

B (continued)

are well exposed on the southeast margin of Mount Ellsworth near the base of the Mount Ellsworth dome (Fig. 146). There, in the Southeast Ellsworth subarea (67 in Fig. 145A), deformation bands were measured in the Entrada Formation. Two systems are evident (Fig. 145B). One system strikes ~N40°W, with conjugate-normal sets dipping both to the northeast and southwest at moderate to steep inclinations. These are cataclastic (DB2c) deformation bands that are marked by slickenlines that rake very steeply. They are aligned *radially* with respect to the Mount Ellsworth igneous center. Deformation bands within this system cross yet another noncataclastic system of deformation bands that lack slickenlines. These deformation bands strike east-northeast and dip steeply both to the north-northwest and to the south-southeast (67 in Fig. 145B).

Both the Southwest Ellsworth and Northwest Ellsworth subareas (68 and 69 in Fig. 145A) reside within the sphere of structural influence of the Mount Ellsworth laccolith (Fig. 147). In the Southwest Ellsworth subarea, the apparent dip of Entrada Formation ranges from ~6° to 15°SW. Deformation bands in the Entrada Formation appear to be noncataclastic in nature. The common range of thickness of deformation bands is from 1 mm to ~10 cm (i.e., up to 4 in.), with most at the several millimeter scale. Normal offsets are the rule, although one reverse-offset deformation band was noted. One really well-developed ladder-structure deformation-band shear zone (DB1c) was noted, and it displayed 10 cm (~4 in.) of normal offset. The deformation bands appear to become thicker and much more densely developed toward the northeast, where the apparent dip increases to ~15°SW. Some of the deformation bands are in fact as thick as 35 cm (14 in.), and these appear to be cataclastic. Stereographic projection of the measured orientations of deformation bands in the Southwest Ellsworth subarea (68 in Fig. 145A) reveals one dominant conjugate-normal system, striking ~N60°E, with sets dipping moderately steeply both southeastward and northwestward. The dominant set is radially oriented with respect to the Mount Ellsworth igneous center. In the Northwest Ellsworth

Figure 146. Nature of the landscape on the southeast flank of Mount Ellsworth. Bedrock is Navajo Sandstone. Photograph taken from outcrops of Entrada Formation. View is to the north.

Figure 147. Photograph of Mount Ellsworth. Note gently dipping Navajo Sandstone (white) at the base.

subarea (69 in Fig. 145A), deformation bands in the Entrada Formation, noncataclastic in nature, form a system of conjugate-normal deformation bands striking north-northwest and dipping both east-northeast and west-southwest. These too are radial in orientation with respect to the Mount Ellsworth igneous center.

Moving from Mount Ellsworth northward to Mount Holmes (Figs. 145A and 31A), my students and I continued to see evidence for a radial pattern of deformation banding within the Entrada Formation, in this case with respect to the Mount Holmes igneous center. In the Southwest Holmes subarea (70 in Fig. 145), the apparent dip of the Entrada Formation is ~15°W, and the deformation bands for the most part strike east-northeast. In the WNW Holmes subarea (71 in Fig. 145), the Entrada Formation dips ~21°NW, and the normal-slip deformation bands strike N60°W. Within the WNW Holmes subarea there is plenty of evidence for the tectonic origin of the deformation bands, including ladder structure, offsets of more than 10 cm (4 in.), and slickenlined surfaces. In the North Holmes subarea (72 in Fig. 145), the deformation bands occur within the upper Navajo Sandstone, and they are very well organized as conjugate-normal faults that strike essentially north. Riedel geometries are evident, and some of the normal-slip displacements are as great as 40 cm (17 in.).

Lower Trail Canyon subarea (73 in Fig. 145A) lies well north of the base of Mount Holmes, southeast of Pulpit Arch, which is a secondary and modest structural dome east of the Mount Hillers igneous center. Deformation bands there occur in Entrada Formation, and the orientations are quite systematic. Two sets of deformation bands are evident stereographically (Fig. 145B): one strikes north-south and dips moderately steeply east; the other strikes ~N50°W and dips moderately steeply northeast.

Less than 2 km (1.2 mi) north of the Lower Trail Canyon subarea is Maidenwater Ravine, a deep slot canyon cut into Navajo Sandstone (74 in Fig. 145A). Within this slot canyon, which is located right next to the highway, is an extraordinary exposure of a thrust-slip Riedel deformation-band shear zone of a cataclastic (DB1) nature (Fig. 147). The shear zone strikes ~N10°W and dips ~30°W. It accommodates offset of ~10 cm (4 in.) (Fig. 148) Overstepping Riedel shears are beautifully expressed, and ladder structures abound. It would seem that the thrusting, which is expressed stereographically as a conjugate system striking approximately north (74 in Fig. 145B), was produced by compression on Navajo Sandstone exerted by the emplacement of the Pulpit Arch intrusive lobe. It seems likely that magma, which originated in the Mount Hillers igneous center (Fig. 145A), exerted west-to-east horizontal pressure near the east end of the Pulpit Arch intrusive lobe, thereby causing horizontal contraction. The photographs in Figure 148 reveal that the contractional strain caused activation of slip and shear along bedding lamination in the Navajo Sandstone. The shear is conspicuously expressed in the white bedding-parallel (DB1d) deformation bands. The flexural-slip shear helped to accommodate the kinking that is evident along the shear zone (Fig. 148B).

Of all of the laccoliths examined in this study, Mount Hillers is the most spectacular, not only because of its daunting structural relief, but especially because of the dramatic way in which the Jurassic strata are domed to nearly vertical orientations around its eastern and southern flanks (Fig. 149). Preferred orientations of deformation bands again "radiate" in trend from the igneous center. The Trail Canyon subarea (75 in Fig. 145) is located east-northeast of the Mount Hillers igneous center, and the deformation bands for the most part strike east-northeast and dip moderately to steeply northwestward. At the Ghost Ridge subarea (76 in Fig. 145), Entrada Formation is rich in deformation bands, which strike northwest and dip moderately to steeply both to the northeast and to the southwest. Bedding strikes ~N30°E and dips ~65°SE, and thus the deformation-band shear zones are essentially orthogonal in strike direction to bedding. Ghost Ridge is a spectacular area physiographically (Fig. 150), and structurally there is abundant evidence for grand-scale slickenlines on bedding surfaces that have been upturned to vertical (Fig. 32). The Copper Creek subarea (77 in Fig. 145) and the Hillers subarea (78 in Fig. 145) are along the southern margin of Mount Hillers. Bed-

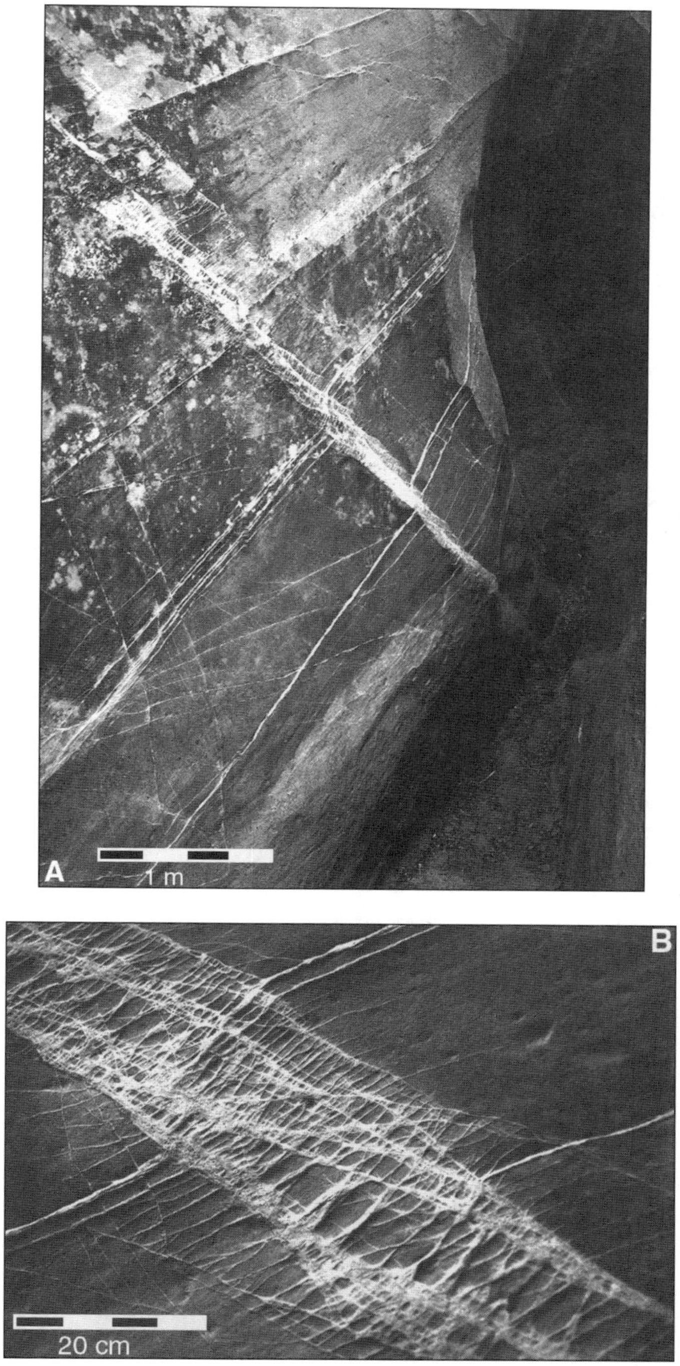

Figure 148. A: West-directed photograph of thrust-slip Riedel deformation-band shear zone, exposed in the slot canyon at Maidenwater Ravine. B: Close-up photograph of the deformation band, revealing thrust offset and Riedel geometries.

ding strikes east and dips at ~80°S. Deformation-band shear zones occur in the Navajo Sandstone, the Entrada Formation, and the Dakota Sandstone. The deformation bands are quite abundant, though most are quite thin, just a matter of several centimeters. The thickest deformation bands are 14 cm (5 in.) thick. Offsets, where evident, are normal or left lateral, and there are abundant deformation bands parallel to cross-lamination. Stereographic projections of the measured deformation bands at the Copper Creek and Hillers subareas (77 and 78 in Fig. 145) show considerable scatter of orientations, but once again the dominant orientations are approximately perpendicular to the strike of bedding and thus radial with respect to the Mount Hillers igneous center.

Origin of the deformation bands and deformation-band shear zones

Results of analysis of deformation bands in the Henry Mountains subregion are generally consistent with the observations and interpretations of Jackson and Pollard (1988a, 1988b, 1990), whose studies were more focused with respect to structural mechanics. Jackson and Pollard (1988) presented elegant maps and cross sections of the fundamental structural characteristics of Mount Ellsworth, Mount Holmes, and Mount Hillers, calling specific attention to the position of zones of peripheral intrusions and faults. Their work emphasizes the expressions of bedding-parallel faulting as an accommodation to flexural shear, as well as the expressions of radial and tangential strain. Their fundamental mechanical model for laccolithic development and doming involves upward deflection of an elastic circular plate clamped around the edges. Their modeling suggests that bending is important especially in the early stages of doming. Then, as doming increases, radial and circumferential extension begin to predominate over bending strains (Jackson and Pollard, 1988a). Jackson and Pollard (1990) emphasized that the formation of deformation bands is one of the principal accommodations to the strains induced by the emplacement of the laccoliths and the formation of the domes.

The record of deformation bands in the Henry Mountains subregion is complicated because of the multiple strain environments that evolved through time, and because early-formed deformation bands (and associated slickenlines) may be rotated to new orientations as the beds within which they occur undergo steepening. The complexity is primarily captured in kinematics: oblique slip is the rule, and the offsets include reverse, normal, left lateral, and right lateral. Because slickenlines are well developed only on a relatively small proportion of the deformation bands, it becomes especially difficult to work out the step-by-step development of the system. My first-order observation is simply this: that the vast majority of deformation bands are radially disposed relative to the respective igneous centers, and they have accommodated normal oblique slip. The radial orientations are a direct structural response to a vertically aligned greatest principal stress direction (σ_1), a least principal stress (σ_3) oriented circumferentially, and an intermediate principal stress (σ_3) oriented radially. It is likely that

Figure 149. A: West-directed view of Mount Hillers, revealing steeply dipping Jurassic strata, notably Navajo Sandstone (white), along the flank. B: Northwest-directed photograph of "tower" of steeply dipping sandstone of the Entrada Formation along the southern margin of Mount Hillers.

Figure 150. Photograph taken westward toward Mount Hillers at Ghost Ridge. Sarah Tindall and Jessica Graybeal are in the foreground.

when the radially oriented deformation bands initially formed, when bedding was subhorizontal, the movements were normal slip. Rotation of bedding then reoriented these deformation-band shear zones (and associated slickenlines) such that they appear as oblique normal–slip deformation bands. Deformation bands accrued throughout the doming process, thus creating the confusing offsetting relationships emphasized by Jackson and Pollard (1990) and apparent in outcrop.

One of the interesting exceptions to the rule pertains to the deformation-band shear zone in Maidenwater Ravine. There, within a compressional environment created by emplacement of the Pulpit Arch intrusive lobe, thrust-slip deformation-band shearing developed within a stress environment marked by a horizontal, east-west orientation of the greatest principal stress (σ_1); a vertical orientation of the least principal stress (σ_3); and a horizontal north-south orientation of the intermediate principal stress (σ_2).

CONCLUSIONS: STRUCTURAL AND TECTONIC DEVELOPMENT OF THE REGION

Results reveal that the Colorado Plateau tectonic province is marked by more internal structural deformation than the prior literature would suggest. In fact, some of the elegance of form and geometry of Colorado Plateau domes and monoclines, where expressed in major sandstone units such as the Navajo Sandstone, is attributable to the complex, intricate, progressively evolving networks and systems of deformation bands that can accommodate smoothly all manner of distortion. The rocks, therefore, disclose a macroscopic ductility (Davis, 1996a). The elegant cross-sectional curvature of Laramide monoclines at the stratigraphic level of the Navajo Sandstone is seen "up close" to have been achieved, in part, by flexural-shear deformation-band formation and by thrust shear along spaced deformation bands, as well as conjugate-normal bending. The elegant sinuosity of Laramide monoclines in plan view is partly accommodated by strike–slip shear and normal-shear deformation-band formation. Similarly, doming of Jurassic strata by emplacement of the mid-Tertiary laccoliths of the Henry Mountains was accommodated, in part, by flexural-shear and circumferential stretching of porous sandstones, products of which are evident in the deformation bands within the Navajo Sandstone and the Entrada Formation. Subregional mid-Tertiary compression, generated by the "toe" effects of the collapse and spreading of the Marysvale volcanic system, is beautifully expressed in thrust-slip deformation-band shear zones within favorable lithologies, notably porous sandstone units within the Straight Cliffs Formation. Even Basin and Range normal faulting left a record of significant deformation-band shearing, notably in footwall Navajo Sandstone along the Sevier fault and in footwall Straight Cliffs Sandstone along the Paunsaugunt fault. Moreover, there are other nontectonic expressions of deformation-band shearing, especially the prelithification, syndiagenetic, compaction-generated deformation-bands that formed within the upper 10 to 15 m (33 to 50 ft) of the Navajo Sandstone, almost wherever it occurs in the region.

Deformation bands appear to form so effortlessly that displacements of any orientation and of any direction can be achieved "on a moments notice" within porous sandstones during any stage of the deformation. I believe that much of the deformation may have been aseismic and commonly favored by the presence of fluids within the porous sandstones. Only when a zone of deformation bands developed to a thickness and stiffness conducive to stick-slip behavior did true faulting take place, presumably with seismic response. The paucity of syn–deformation-band jointing in rocks that are especially strongly deformed by deformation banding suggests that deformation by shear in porous sandstone units may actually require less work than deformation by tensile failure.

Differential stresses on fluid-filled porous sandstone causes collapse of porosity and an advance of a "front" of elevated fluid pressure that escapes beyond the tip of the deformation band and beyond the tip of the process zone (Fig. 57). Where grains are pressed into contact, stress is concentrated to the point that the

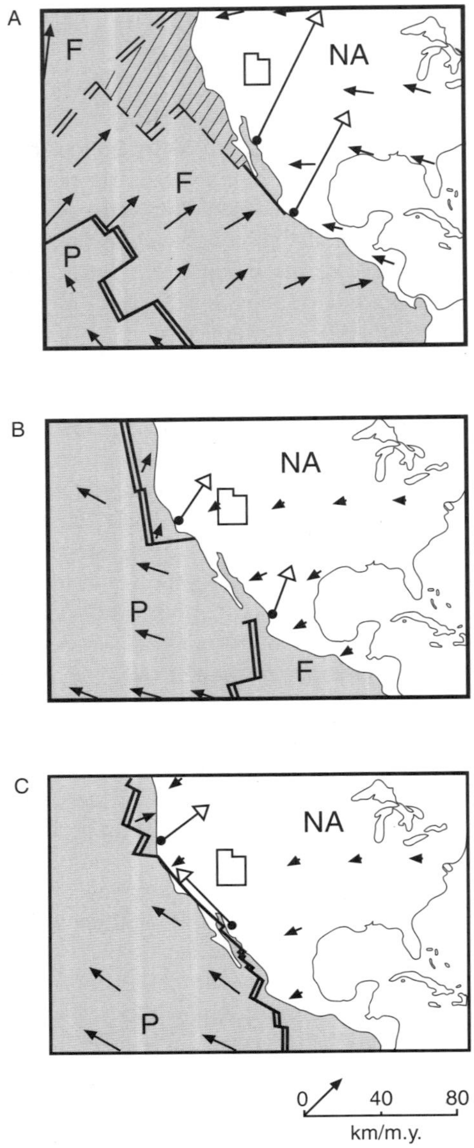

Figure 151. Plate reconstructions for three periods of deformation. A: Laramide. B: Early Miocene. C: Basin and Range. NA = North American plate. P = Pacific plate. F = Farallon plate. Rates of absolute plate motion given by bar scale and solid-arrow-tipped lines. Open-arrow-tipped lines show rates of relative motion between North American and Pacific plates: Vector lengths are ~100 mm/yr for Laramide; ~50 mm/yr for early Miocene; and ~48 mm/yr for Basin and Range. Illustrations adapted from Engebretson et al. (1985).

fracture strength of individual grains is surpassed. Such stress concentration thus leads to grain-scale fracturing and crushing as slip and/or shear is accommodated by the developing deformation-band shear zone. The angularity of broken fragments and the loss of pore space then tends to create work hardening and frictional lock-up. Under conditions of modest differential stress and fluid-pressure buildup, deformation bands may have developed as transitional-tensile "anticrack" shears.

For cataclastic deformation bands of cataclastic origin, the outcrop-scale deformation mechanism was Riedel shearing, commonly conjugate. We see in these patterns the incipient development of fault zones "frozen" in time. Although not exploited in this study, the presence of such varied systems of deformation bands within the southern Utah parks and monuments region creates an opportunity for systematic analysis of the physical characteristics (and mechanics of origin) of deformation-band shear zones as a function of confining pressure. Diagenetic compaction bands within uppermost Navajo Sandstone formed in the Jurassic at depths of less than 0.5 km. Thrust-slip deformation bands within the Straight Cliffs Formation formed in the mid-Tertiary at depths of ~1.5 km (1 mi) or less. Normal-slip deformation-band shear zones formed in Navajo Sandstone during Basin and Range deformation at depths of 2 km (1.3 mi) or less. Deformation bands of flexural-slip, normal-slip, thrust-slip, and strike-slip shear origin formed within Navajo Sandstone at depths of 3 to 4 km (1.8 to 2.4 mi) along the Laramide monoclines and the mid-Tertiary laccoliths of the Henry Mountains.

It is useful to place the development of Colorado Plateau structures, as well as the formation of deformation-band shear zones within the region, into a plate-tectonic and regional-tectonic context. In doing so, I present plate-tectonic reconstructions (Fig. 151) for Laramide, early Miocene, and Basin and Range deformation episodes; and I invert the basic geometric and kinematic signatures of subarea clusters of deformation bands into paleo–focal mechanisms (Fig. 152). Inversion to create paleo–focal mechanisms is carried out with full knowledge that not all of the deformation-band shear zones were produced coseismically, for it is possible that some of this deformation was achieved through aseismic creep. Yet the paleo–focal mechanisms are a useful integrative image that can help summarize the total kinematic significance of the subarea swarms of deformation-band shear-zone data.

When Navajo Sandstone was being deposited in southern Utah, southwestern North America was located at lat ~10°N and existed as a plate margin distinguished by steep subduction of oceanic lithosphere (Dickinson, 1989; Tosdal et al., 1989; Saleeby et al., 1992). The direction of convergence may have been about N40°E, though the rate of relative motion between North America and oceanic plate(s) to the west is unknown. The absolute motion of North America was ~45 mm/yr. Subduction of oceanic lithosphere was weakly coupled with the overriding plate, and southwestern North America may have been undergoing weak extension. Results of this study suggest that the Navajo Sandstone, during burial, was compacted and flattened in a process aided by pore fluids and accommodated by the development of pervasive nontectonic, noncataclastic deformation-band shear zones in the upper 15 m (50 ft) of the formation. These structures developed pervasively through granular-flow mechanisms. At the outcrop scale, these deformation bands are distinguished by con-

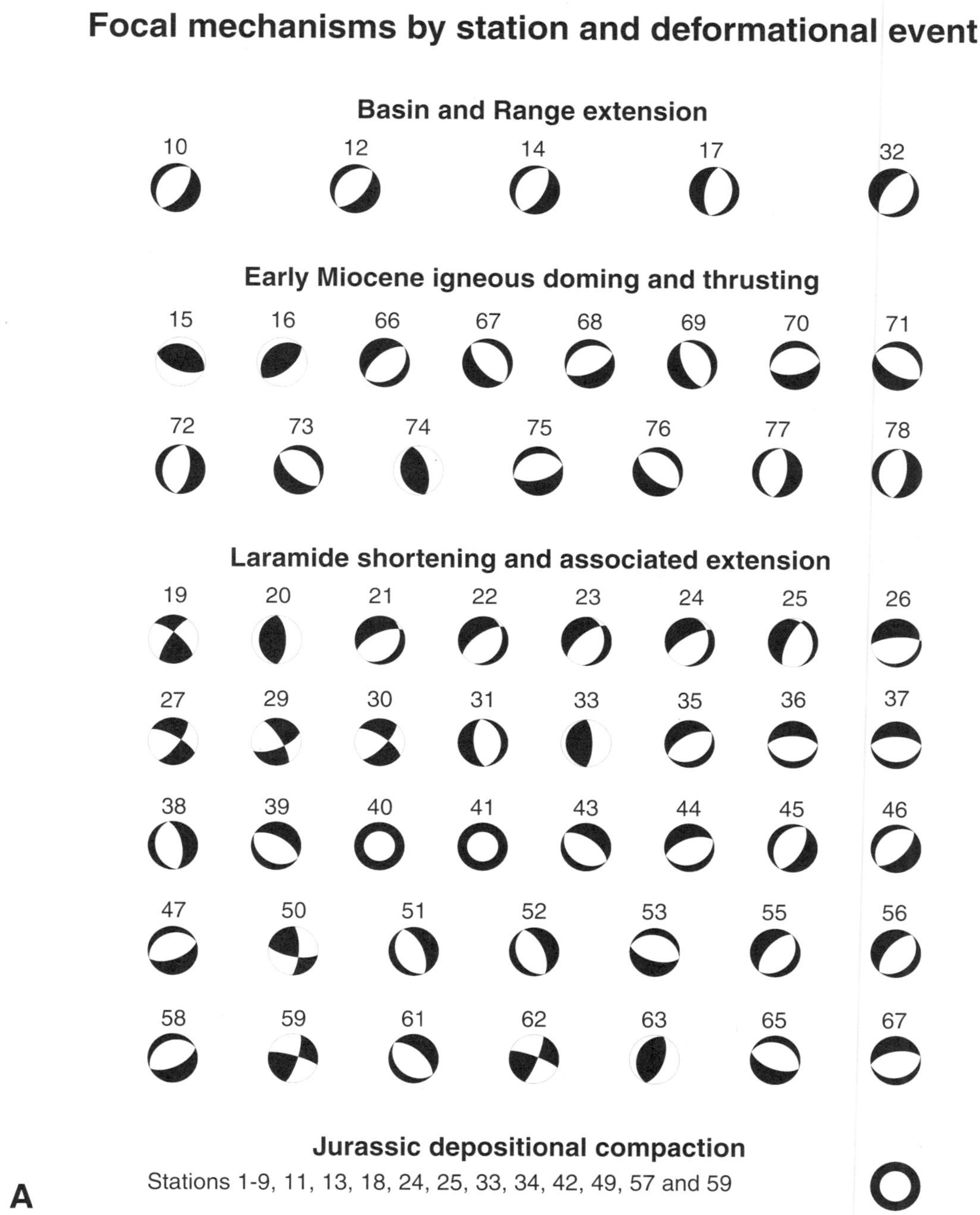

Figure 152 (this and following four pages). Paleo–focal mechanisms inverted from deformation data, by station. A: Station-by-station paleo–focal mechanisms inverted from deformation-band data. B: Map of Laramide paleo–focal mechanisms. C: Map of early Miocene and Basin and Range paleo–focal mechanisms.

jugate-normal shear zones with strike orientations that completely box the compass. Dip inclinations average ~60°, though listric geometries are common.

During the Laramide, and at ca. 70 Ma, Utah was located at approximately lat 40°N. The plate boundary along the western edge of the North American plate was marked by low-angle subduction of oceanic lithosphere (Severinghaus and Atwater, 1990). The direction of convergence between North America and the Farallon plate was ~N30°E (Fig. 151A) (Engebretson et al., 1985; Severinghaus and Atwater, 1990), though it may have been

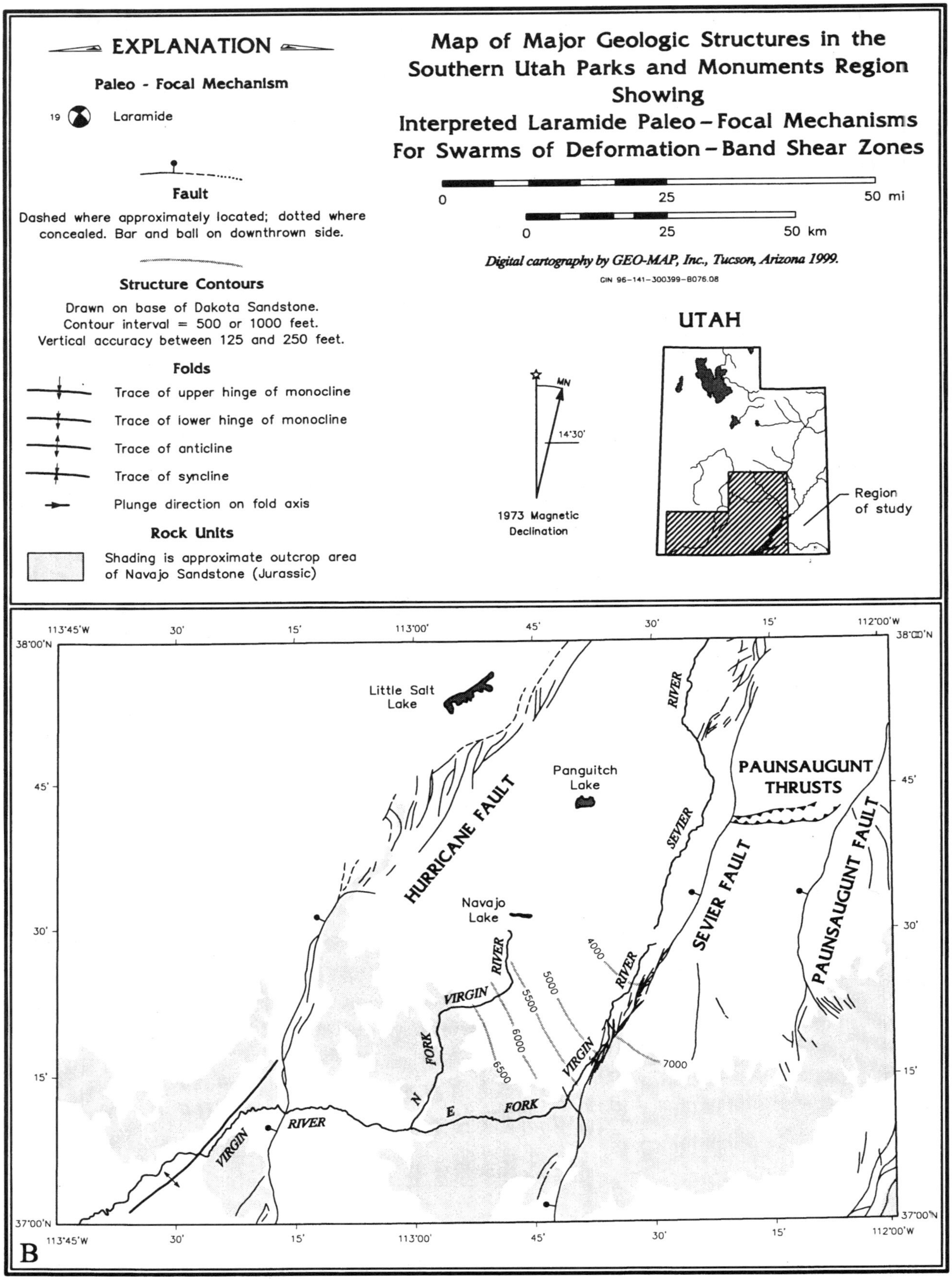
EXPLANATION
Paleo - Focal Mechanism
19 Laramide
Fault
Dashed where approximately located; dotted where concealed. Bar and ball on downthrown side.
Structure Contours
Drawn on base of Dakota Sandstone.
Contour interval = 500 or 1000 feet.
Vertical accuracy between 125 and 250 feet.
Folds
Trace of upper hinge of monocline
Trace of lower hinge of monocline
Trace of anticline
Trace of syncline
Plunge direction on fold axis
Rock Units
Shading is approximate outcrop area of Navajo Sandstone (Jurassic)
Map of Major Geologic Structures in the Southern Utah Parks and Monuments Region Showing Interpreted Laramide Paleo-Focal Mechanisms For Swarms of Deformation-Band Shear Zones
0
25
50 mi
0
25
50 km
Digital cartography by GEO-MAP, Inc., Tucson, Arizona 1999.
GIN 96-141-300399-B076.08
UTAH
MN
14°30'
1973 Magnetic Declination
Region of study
113°45'W
30'
15'
113°00'
45'
30'
15'
112°00'W
38°00'N
45'
30'
15'
37°00'N
Little Salt Lake
Panguitch Lake
Navajo Lake
HURRICANE FAULT
PAUNSAUGUNT THRUSTS
SEVIER FAULT
PAUNSAUGUNT FAULT
SEVIER
RIVER
VIRGIN
RIVER
N FORK
E FORK
4000
5000
5500
6000
6500
7000
B

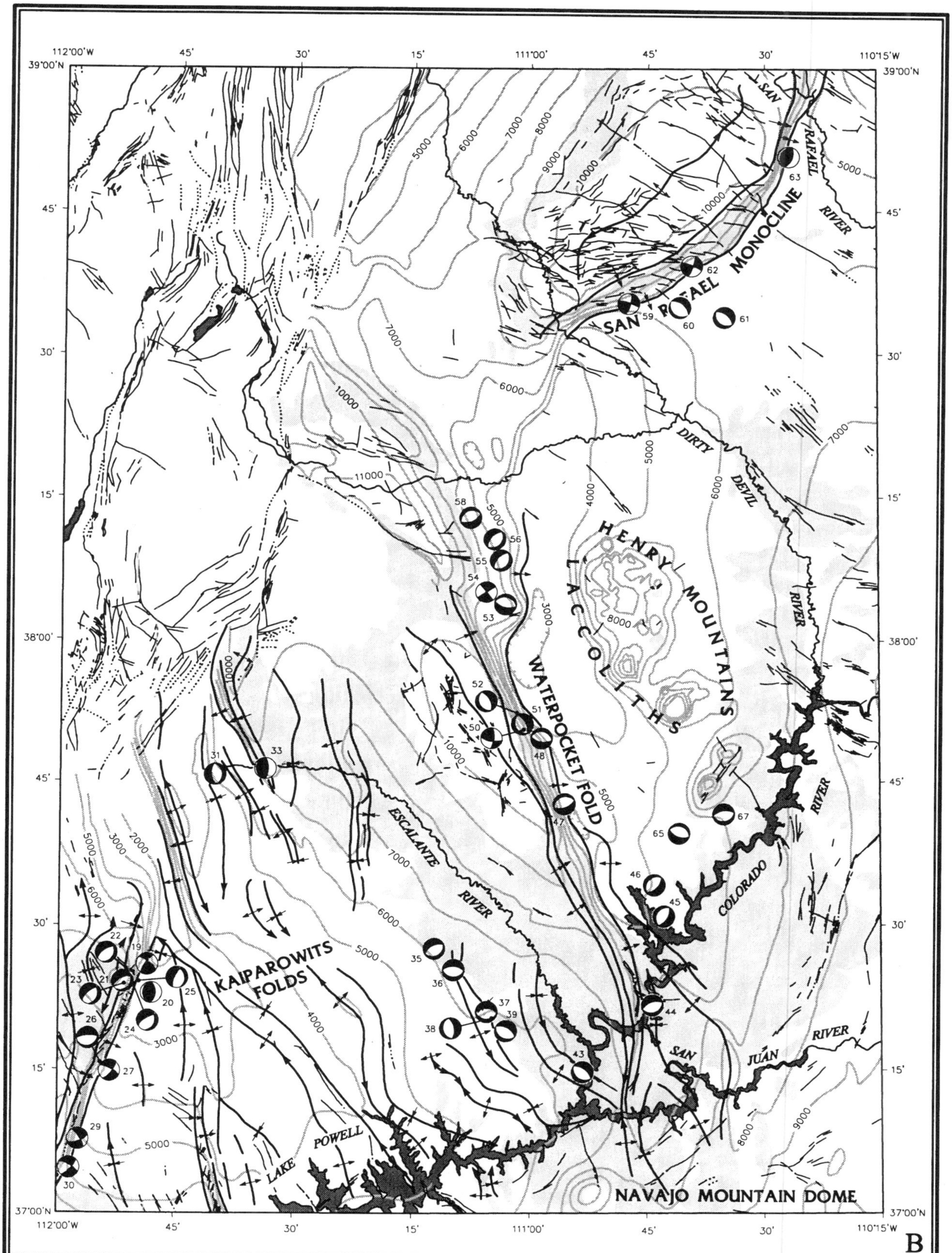
112°00'W
45'
30'
15'
111°00'
110°15'W
39°00'N
37°00'N
38°00'
SAN RAFAEL MONOCLINE
SAN RAFAEL RIVER
DIRTY DEVIL RIVER
HENRY MOUNTAINS LACCOLITHS
WATERPOCKET FOLD
ESCALANTE RIVER
COLORADO RIVER
KAIPAROWITS FOLDS
SAN JUAN RIVER
LAKE POWELL
NAVAJO MOUNTAIN DOME
B

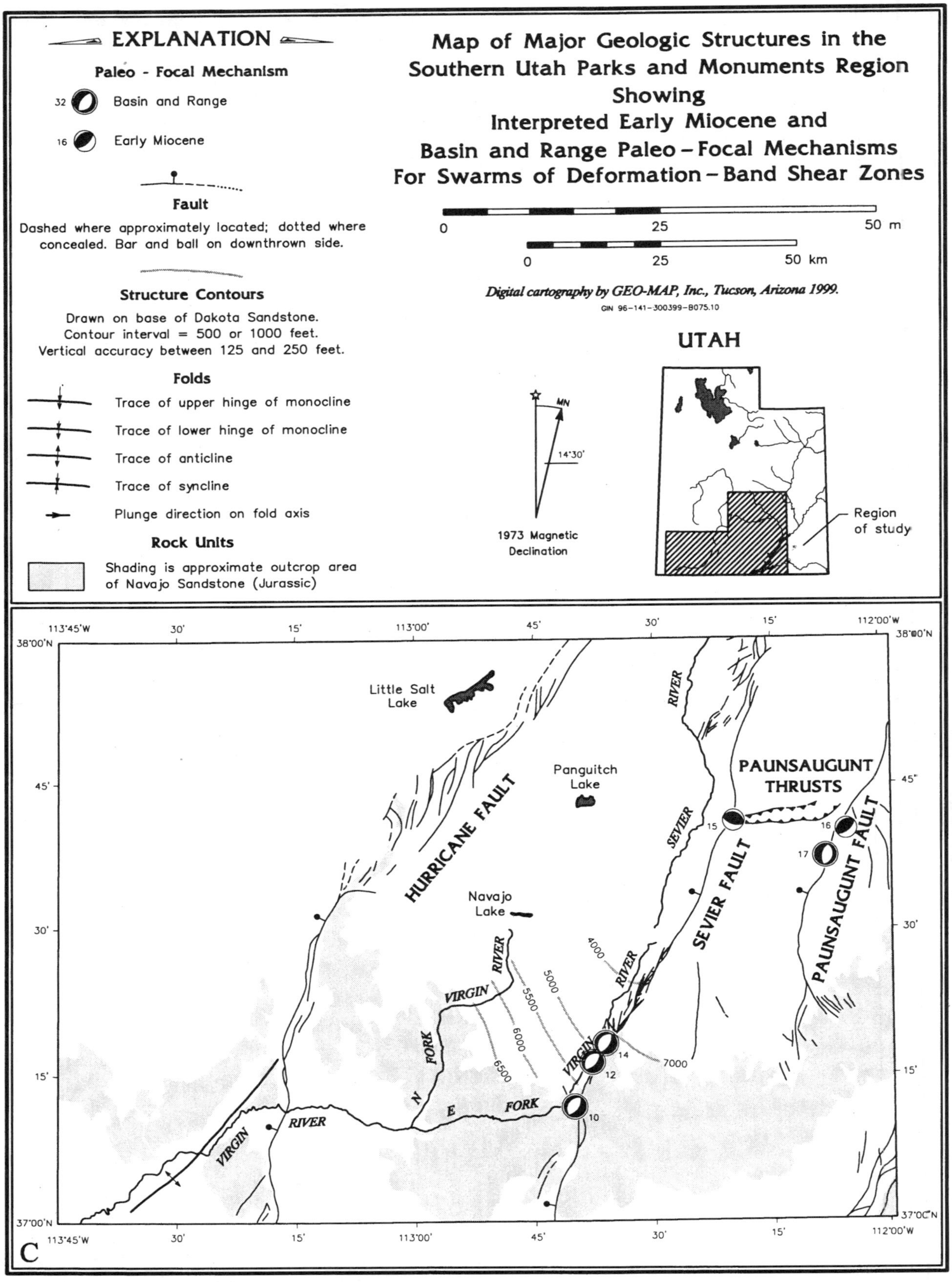
EXPLANATION
Paleo - Focal Mechanism
32 Basin and Range
16 Early Miocene
Fault
Dashed where approximately located; dotted where concealed. Bar and ball on downthrown side.
Structure Contours
Drawn on base of Dakota Sandstone. Contour interval = 500 or 1000 feet. Vertical accuracy between 125 and 250 feet.
Folds
Trace of upper hinge of monocline
Trace of lower hinge of monocline
Trace of anticline
Trace of syncline
Plunge direction on fold axis
Rock Units
Shading is approximate outcrop area of Navajo Sandstone (Jurassic)
Map of Major Geologic Structures in the Southern Utah Parks and Monuments Region Showing Interpreted Early Miocene and Basin and Range Paleo-Focal Mechanisms For Swarms of Deformation-Band Shear Zones
0 25 50 m
0 25 50 km
Digital cartography by GEO-MAP, Inc., Tucson, Arizona 1999.
GIN 96-141-300399-8075.10
UTAH
MN
14°30'
1973 Magnetic Declination
Region of study
113°45'W 30' 15' 113°00' 45' 30' 15' 112°00'W
38°00'N 45' 30' 15' 37°00'N
Little Salt Lake
Panguitch Lake
Navajo Lake
HURRICANE FAULT
SEVIER FAULT
SEVIER RIVER
PAUNSAUGUNT THRUSTS
PAUNSAUGUNT FAULT
VIRGIN RIVER
N FORK VIRGIN RIVER
E FORK VIRGIN
4000 5000 5500 6000 6500 7000
10 12 14 15 16 17
C

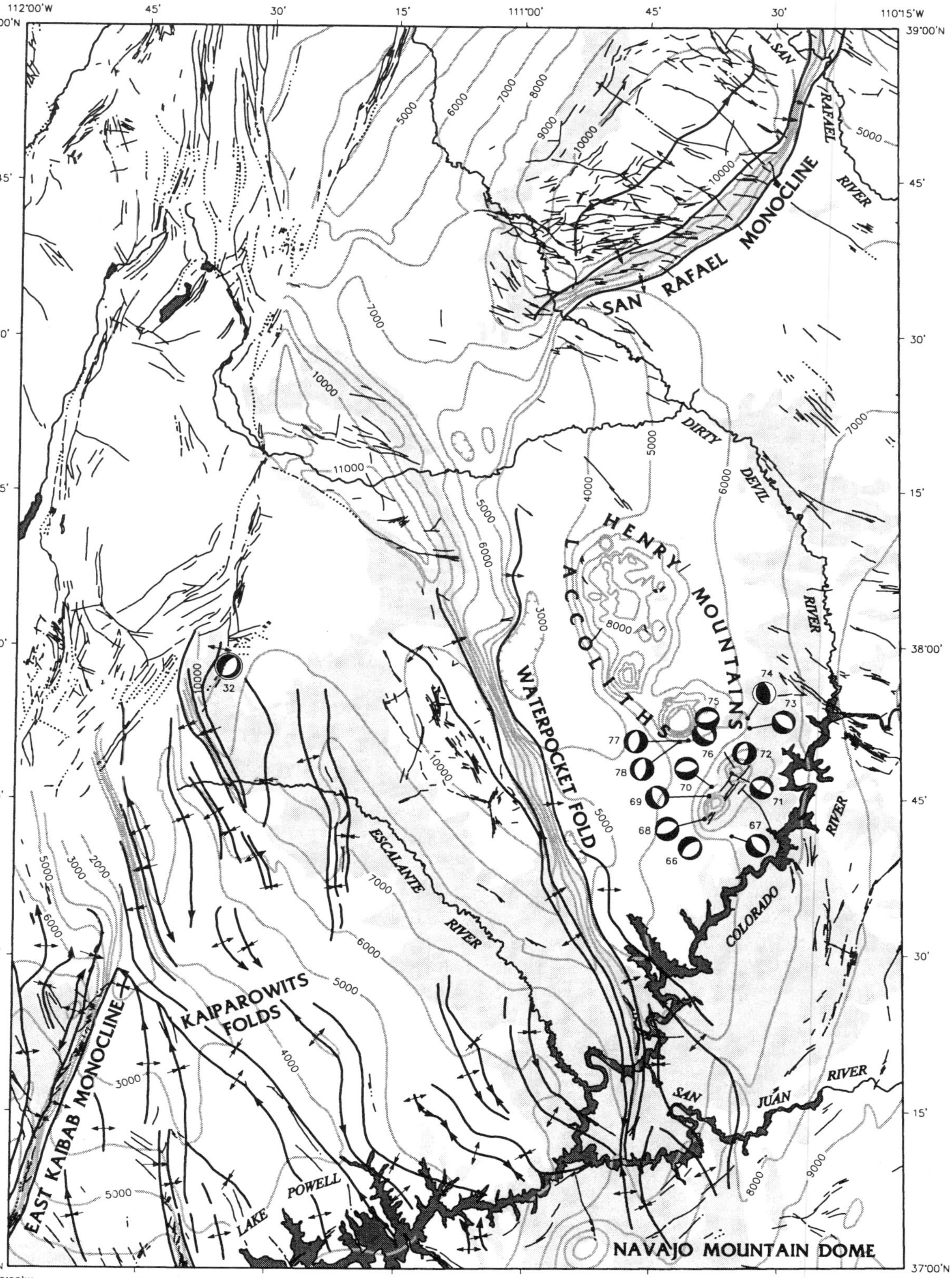

SAN RAFAEL MONOCLINE
SAN RAFAEL RIVER
DIRTY DEVIL RIVER
HENRY MOUNTAINS LACCOLITHS
WATERPOCKET FOLD
ESCALANTE RIVER
COLORADO RIVER
KAIPAROWITS FOLDS
EAST KAIBAB MONOCLINE
LAKE POWELL
SAN JUAN RIVER
NAVAJO MOUNTAIN DOME
112°00'W
111°00'
110°15'W
39°00'N
38°00'
37°00'N

C

as far toward the east as N60°E (Coney, 1978). The convergence rate is estimated at 100 mm/yr (Coney, 1978; Stock and Molnar, 1988; Engebretson et al., 1985). Absolute motion of North America was 10 mm/yr, oriented ~S30°W. It was during the Laramide that basement-cored uplifts developed in the foreland of the Cordilleran orogenic belt, including the monoclines within the southern Utah parks and monuments region. Northeast-southwest compressional stresses created the major uplifts, including the Kaibab, Circle Cliffs, Miners Mountain, and San Rafael uplifts. Each of these uplifts is associated with a major monocline: the East Kaibab, Waterpocket, Miners Mountain, and San Rafael. The Circle Cliffs uplift and Waterpocket Fold are oriented essentially perpendicular to the Laramide regional compressive stress, and thus it may be inferred that they were formed by pure thrust and/or reverse faulting, i.e., faulting with no significant strike-slip component. This inference is also true of the north-northwest–striking, southwest-facing Escalante monocline. In contrast, the Kaibab uplift and East Kaibab monocline appear to be right-handed–reverse transpressional structures, and the Miners Mountain uplift and associated monocline appear to be left-handed–reverse transpressional structures. Alex Bump (personal commun., 1999) has concluded that the San Rafael Swell and monocline formed under the influence of west-northwest–east-southeast contraction, i.e., parallel to the Sevier orogeny direction of contraction that commenced in Cretaceous time. Thus the San Rafael Swell may have formed relatively early in comparison to the other Colorado Plateau basement-cored uplifts.

The Laramide folding and faulting created a myriad of cataclastic, tectonic, deformation-band shear zones within the Navajo Sandstone, and these overprint the noncataclastic deformation bands of compactional origin that reside in the uppermost Navajo Sandstone. The Laramide deformation-band shear zones are broadly distributed in Navajo Sandstone across the full width and length of each of the monoclinal folds (Fig. 152, A and B), but they are essentially absent in the flat-lying strata on either side of the folds proper. The varieties of deformation bands attributable to Laramide origin can be seen in Appendix Table 2. Some were created by extension in the outer arc of folds during bending and thus are seen in the extensional hinge regions of monoclines, anticlines, and synclines. Some deformation bands are layer parallel and were created by flexural-slip shear during folding, especially as a result of the out-of-the-syncline shearing required by the space problem in the lower hinge regions of monoclines. Some deformation bands record contraction and shortening within the lower parts of the limbs of monoclines and are expressed as thrust- and reverse-slip shear zones. Some deformation bands are synthetic to master faults that "drove" the fault-propagation folding (now expressed as monoclines) and eventually broke through the folds. The linkage between folding and faulting may be trishear in nature (Erslev, 1991). Some deformation bands are purely strike slip, permitting modest shortening perpendicular to the trace(s) of monoclines and modest stretching parallel to them (e.g., the Burr Trail area of the Waterpocket Fold). Finally, the widespread deformation banding evident in the Entrada Formation may reflect a modest extension at right angles to the direction of Laramide shortening (Fig. 152A).

The deformation banding within monoclines represents a *complete* record of the entire progressive strain during folding, and it is for this reason that the patterns are so complex. Yet the complexity is interpretable, and it is usually possible to gain a reliable sense of whether the kinematics of folding were predominantly dip slip, strike slip, or oblique slip and whether local strain induced stretching or shortening.

In the early Miocene (at ca. 20 Ma), Utah was located at approximately lat 40°N. To the west, the North American and Farallon plates, at latitudes north of the Mendocino triple junction (which was at lat 32°N), were on a N30°E–S30°W convergent course at a rate of ~50 mm/yr (Fig. 151B). The oceanic lithosphere, which had been subducting at a relatively shallow angle, was rolling back to a steeper subduction angle. Early Miocene magmatism induced by subduction was well developed in the foreland region, notably in the Marysvale volcanic field of southwestern Utah, in the Henry Mountains of southeastern Utah, and at Navajo Mountain. Igneous activity "drove" structural deformation in Utah at this time. Emplacement of the laccoliths of the Henry Mountains resulted in the doming of the full Jurassic and Cretaceous stratigraphic columns at five discrete centers. As the strata were intruded, they were steepened to inclinations as great as 85°. The doming and uptilting of strata resulted in flexural-slip shearing along bedding and radial and circumferential stretching. Both of these structural mechanisms left a record of normal-oblique-slip deformation-band shear zones within porous sandstones, notably the Navajo Sandstone and the Entrada Formation (Fig. 152C). Less commonly, magma lobes that migrated horizontally away from the igneous centers locally created compression that caused development of deformation-band shear zones of a reverse or thrust nature. The difficulty in interpreting the origin of the deformation bands in the vicinity of the laccoliths is partly due to progressive deformation and rotation. Early-formed deformation bands of a dip-slip extensional nature have become rotated as the strata within which they occur became tilted to steeper and steeper angles. Thus, deformation-band shear zones oriented radially with respect to an igneous center may have been marked originally by 90°-raking slickenlines. Yet, as the beds and shear zones were rotated to steeper angles, the slickenlines have assumed progressively lower rake angles and thus give the false impression, in many cases, that the kinematics were oblique slip or strike slip. Furthermore, new deformation-band shear zones formed progressively during the igneous doming, affecting bedding that was relatively low dipping early in the emplacement history, relatively steeply dipping later in the emplacement history, and everything in between. Working out the early Miocene deformation-band history in the Henry Mountains is most difficult where the host rock is Entrada Formation, for the Entrada already contained a record of Laramide deformation banding.

Some early Miocene deformation was of a compressional nature, expressed in the form of the Paunsaugunt thrust belt and associated thrust-slip deformation-band shear zones (Fig. 152C). These structures are found along a 100 km (66 mi) belt and thus are subregional in nature. They represent the "toe" effect of a massive collapse and spreading of the southern part of the Marysvale volcanic field. Collapse and spreading was aided by failure on the Carmel Formation evaporites (Jurassic), for thrusts sole into this unit. Thrusts moved Cretaceous strata over Eocene strata, and some of the deformation was accommodated by fault-propagation folding. Deformation banding produced by the thrusting occurs in porous sandstone units of the Cretaceous Straight Cliffs Formation. This signature of deformation is broadly distributed, occurring over a distance of at least several kilometers in the direction of tectonic transport, within both hanging-wall and footwall strata. This record indicates that even the subtlest record of deformation-band shear zones may attest to the presence of a local, perhaps significant, tectonic structure.

During middle to late Tertiary time, Utah underwent extensional deformation, especially in the western half of the state in the Basin and Range province (Fig. 151C). In southern Utah, the Basin and Range structures "bite into" the Colorado Plateau in the form of the Sevier and Paunsaugunt faults. The Paunsaugunt fault is the easternmost Basin and Range fault in southern Utah, and its location may have been predetermined by the easternmost extent of evaporites in the Carmel Formation, creating a mechanical control (Frank Royse, personal commun., 1986). From a plate-tectonic perspective, the absolute motion of the North American plate at 10 Ma was ~S60°W at 10 mm/yr. To the north of the Mendocino triple junction (which at this time was positioned at lat ~35°N), North America and the Juan de Fuca plate converged along an ~N40°E line of movement at ~45 mm/yr, while to the south of the triple junction the San Andreas transform accommodated right-lateral strike slip at a rate of nearly 48 mm/yr. Extension in the southern Utah parks and monuments region produced the Hurricane, Sevier, and Paunsaugunt faults and related splays. In the Navajo Sandstone, these zones of faulting are marked by deformation-band shear zones of a normal-slip nature (Fig. 152C). They seldom extend more than 50 m (155 ft) into the footwall of the faults, and thus the deformation is more highly localized than the record of deformation-band shear zones that is found in compressional settings. The normal-fault deformation-band shear zones for the most part are oriented parallel to the major fault, but secondary sets are orthogonal to the main trend, suggesting that the intermediate principal stress direction (σ_2) and the least principal stress direction (σ_3) swapped positions during progressive deformation.

Results have clarified the nature and extent of intraplate tectonics withing the Colorado Plateau. Commonly portrayed as tectonically benign, the Colorado Plateau is actually quite interestingly and substantively deformed (Fig. 153). Intraplate regions have been shortened, shifted, stretched, and domed with respect to one another, creating challenging configurations of transfer, interference, and superposition.

In conclusion, the southern Utah parks and monuments region is a valuable natural laboratory for the study of faults and fault-related structures. The deformation-band shear zones stand out resistantly to erosion, and their interiors are commonly winnowed. Consequently, the internal architecture of this type of fault structure is laid bare, allowing examination of the full three-dimensional properties of these shear structures. This natural laboratory of deformation bands will be a special place for evaluating whole shear zones, from tip to tip, affording insight regarding why fault zones begin and end where they do. Of course, the deformation-band shear zones impart a strong permeability anisotropy to the host rock. The zones are marked by negligible porosity and very limited permeability. Thus, they act as barriers to fluid flow. In this regard, the net permeability flux created by the formation of deformation banding is of critical significance in understanding fluid flow within reservoir rocks. All and all, it is my hope that this overview of some of the exquisite internal structure of rocks in the Colorado Plateau may stimulate additional studies, including microscopic fabric analyses, that will generate new knowledge in the understanding of deformation of the Earth's upper crust.

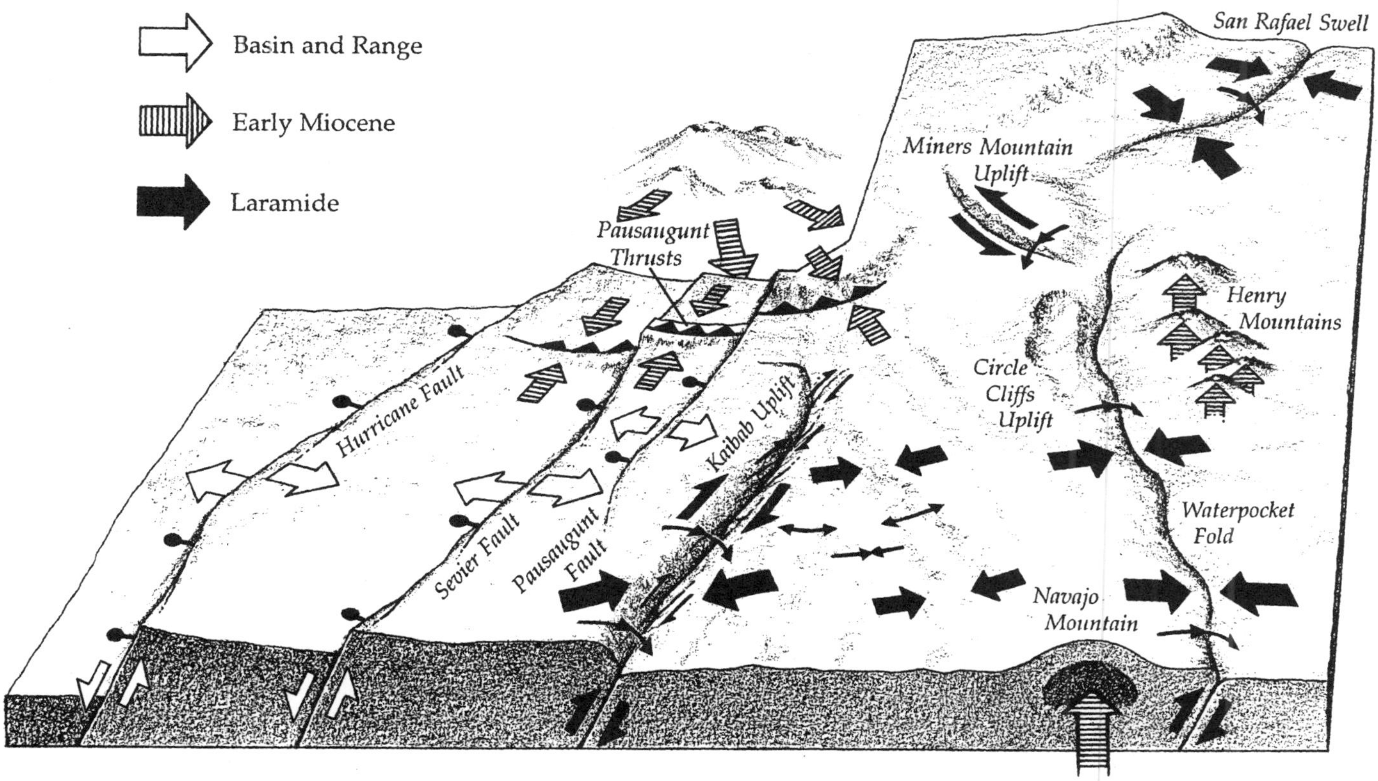

Figure 153. Schematic rendering of intraplate deformations within the Colorado Plateau region of southern Utah. Artwork by David A. Fischer.

Cataclastic varieties (DB1): distinguished on the microscopic scale by conspicuous *grain-scale fracturing and cataclasis* as well as *collapse of pore space and volume reduction* (Antonellini, Aydin, and Pollard, 1994).

Text Figure 97B.

Subvariety DB1a: distinguished in outcrop by *parallel, discrete zones* of deformation bands and zones of deformation bands. Associated commonly with *slickensides and slickenlines* along margins. *Offsets readily apparent,* ranging from millimeters across deformation bands to centimeters or even meters across zones of deformation bands.

Text Figure 60A.

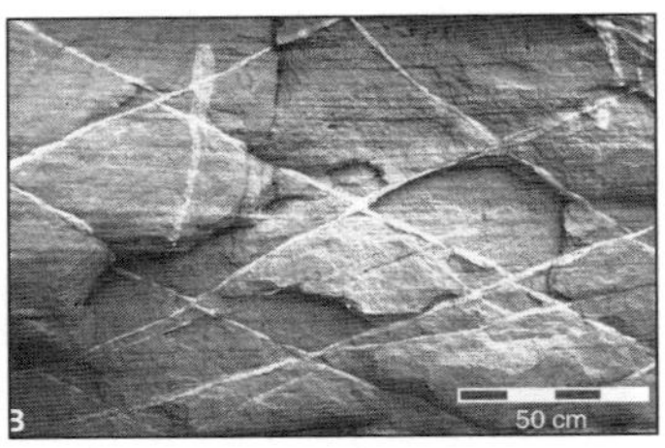

Subvariety DB1b: distinguished in outcrop by *intersecting sets of conjugate* deformation bands and zones of deformation bands. Associated commonly with *slickensides and slickenlines* along margins. *Offsets readily apparent,* ranging from millimeters across deformation bands to centimeters or even meters across zones of deformation bands.

Text Figure 60B.

Subvariety DB1c: distinguished in outcrop by *ladder structure,* especially well developed in areas between overstepping deformation bands and zones of deformation bands. Associated commonly with *slickensides and slickenlines* along margins. *Offsets readily apparent,* ranging from millimeters across deformation bands to centimeters or even meters across zones of deformation bands.

Text Figure 60C.

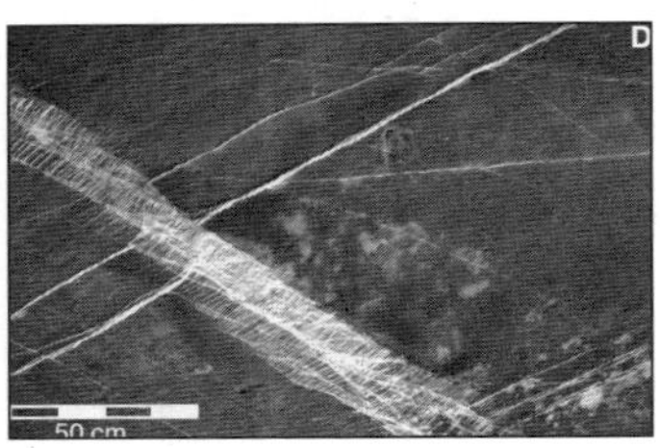

Subvariety DB1d: distinguished in outcrop by *bedding-parallel* deformation bands and zones of deformation bands (in figure, white lines dipping right to left). Associated commonly with *slickensides and slickenlines* along margins. *Offsets readily apparent,* ranging from millimeters across deformation bands to centimeters or even meters across zones of deformation bands.

Text Figure 60D.

APPENDIX TABLE 1. (continued)

Noncataclastic varieties (DB2): may have *some grain-scale fracturing* but marked by *little or no cataclasis;* can form with dilatancy, with compaction, or with no volume change (Antonellini, Aydin, and Pollard, 1994).

Text Figure 71C.

Subvariety DB2a: distinguished in outcrop by *very thin, very closely spaced, wispy, irregularly curviplanar* deformation bands that occur in *multiple conjugate sets.* Millimeter scale in thickness, and centimeter to meter scale in trace lengths. *Absence of slickenlines and slickensides. Negligible offsets* observed across deformation bands, generally less than several millimeters. *Found in study area only in uppermost 10 to 15 m of Navajo Sandstone.*

Text Figure 62A.

Subvariety DB2b: distinguished in outcrop by *long, continuous, well-defined conjugate zones,* meter scale in spacing and length, and *centimeter scale in thickness. Offsets commonly negligible,* but can be as great as several centimeters. *Found in study area only in uppermost 10 to 15 m of Navajo Sandstone.*

Text Figure 63.

Subvariety DB2c: distinguished in outcrop by *long, continuous, thin* (as much as several millimeters) *to moderately thick* (as much as 30 cm), individual, *discrete planar to curviplanar* bands. Occur in *systems of subparallel to anastomosing and inosculating* deformation bands that are *spaced at the meter scale.* Generally *no more resistant to erosion than host rock. Found in study area only in Entrada Formation.*

Text Figure 65A.

Subvariety DB2d: distinguished in outcrop by relatively *closely spaced (10 to 20 cm), thin, irregular but relatively continuous* deformation bands along which *small fault offsets* (centimeter scale) are commonly evident. Seldom do these project in topographic relief. Commonly *bleached white or lighter (or darker) rose-colored. Found in study area only in Entrada Formation.*

Text Figure 66A.

Pressure-solution variety (DB3): "compaction" bands that *form perpendicular to tectonic contraction* and that accommodate shortening by *collapse of pore space* and by *grain-contact pressure solution* (Mollema and Antonellini, 1996).

Variety DB3: Distinguished in outcrop by *relatively widely spaced* (centimeter to meter scale) *planar and continuous* zones as much as 10 m in trace length, to *stylolitic, short, and crooked* deformation bands with very short trace lengths (only 1 or 2 m). *No evidence for shear offset* across these structures. Not observed in study area, but perhaps overlooked (?).

APPENDIX TABLE 2. DATA AND INTERPRETATIONS

	A	B	C	D	E
1	SUBREGION/SUBAREA	TYPE OF BANDS	ROCK FORMATION	DOMINANT ORIENTATIONS	SLICKENLINE ORIENTATIONS
2	#1. ZION/UPPER KOLOB PLATEAU	DB2a	UPPERMOST NAVAJO	BALL OF STRING, DIP ~30-75	NONE
3	#2. ZION/UPPER KOLOB CREEK	DB2a	UPPERMOST NAVAJO	BALL OF STRING, DIP ~30-75	NONE
4	#3. ZION/KOLOB TERRACE	DB2a	UPPERMOST NAVAJO	BALL OF STRING, DIP ~30-75	NONE
5	#4. ZION/ZION EAST ENTRANCE	DB2a	UPPERMOST NAVAJO	BALL OF STRING, DIP ~30-75	NONE
6	#5. ZION/WHITE CLIFFS	DB2a	UPPERMOST NAVAJO	BALL OF STRING, DIP ~30-75	NONE
7					
8					
9	#6. SEVIER/YELLOWJACKET BLUFF	DB2a	UPPERMOST NAVAJO	BALL OF STRING, DIP ~30-75	VERY RARE
10	#7. SEVIER/MONUMENT KNOLL	DB2b	UPPERMOST NAVAJO	N60W STRIKE, DIP ~65-75SW	RAKE ~70-80W
11		DB2b	UPPERMOST NAVAJO	NS STRIKE, DIP ~45-55E	NONE
12	#8. SEVIER/JOHNSON CANYON	DB2b	UPPERMOST NAVAJO	EW STRIKE, DIP ~60-70N	NONE
13		DB2b	UPPERMOST NAVAJO	EW STRIKE, DIP ~60-70S	NONE
14	#9. SEVIER/DIANA'S THRONE	DB2b	UPPERMOST NAVAJO	N60W STRIKE, DIP ~60-70NW	RARE, RAKE ~75W
15		DB2b	UPPERMOST NAVAJO	N60W STRIKE, DIP ~60-70 SW	RARE, RAKE ~70E
16	#10. SEVIER/SLOT CANYON	DB1a	NAVAJO	N30-40E STRIKE, DIP ~60-80NW	RAKE ~80SW
17		DB1a	NAVAJO	N30E STRIKE, DIP ~65SE	NONE
18	#11. SEVIER/MOUNT CARMEL JCT	DB2b	UPPERMOST NAVAJO	BALL OF STRING, DIP ~40-80	NONE
19	#12. SEVIER/ORDERVILLE	DB1a	NAVAJO	N40E STRIKE, DIP ~50-85SE	NONE MEASURED
20		DB1a	NAVAJO	N45E STRIKE, DIP ~75-85NW	NONE MEASURED
21	#13. SEVIER/RED HOLLOW	DB2a	UPPERMOST NAVAJO	BALL OF STRING, DIP ~60-80	NONE
22	#14. SEVIER/GLENDALE	DB1a	NAVAJO	EW STRIKE, DIP ~70-85N & S	NONE MEASURED
23		DB1a	NAVAJO	N30E STRIKE, DIP ~70-85SE & NW	NONE MEASURED
24					
25					
26	#15. PAUNSAUGUNT/HILLSDALE CANYON	DB1abcd	STRAIGHT CLIFFS	N70W STRIKE, DIP ~20-35NE & SW	RAKE ~80-90
27	#16. PAUNSAUGUNT/TROPIC DITCH	DB1ab	STRAIGHT CLIFFS	N85W STRIKE, DIP ~15-25N	NONE
28		DB1ab	STRAIGHT CLIFFS	N50E STRIKE, DIP ~30-40SE	NONE
29	#17. PAUNSAUGUNT/SINKING SHIP	DB1b	STRAIGHT CLIFFS	N10E STRIKE, DIP ~70SE & NW	RAKE ~70-90N
30					
31					
32	#18. KAIBAB/LOWER SLICKROCK	DB2a	PAGE, in CARMEL	N20W STRIKE, DIP ~55-85	RARE, RAKE ~70-90
33	#19. KAIBAB/CAD'S CROTCH	DB1abc	STRAIGHT CLIFFS	NW STRIKE, NE DIP	RAKE ~10-20E
34		DB1abc	STRAIGHT CLIFFS	ENE STRIKE, NNW DIP	RAKE ~10-20E
35		DB1d	STRAIGHT CLIFFS	NS STRIKE, DIP ~30-70E	NONE
36	#20. KAIBAB/LUNCH SPRING	DB1b	WAHWEAP	NS STRIKE, DIP ~20-30W	NONE
37		DB1b	WAHWEAP	N20W, DIP ~ 25-30E	NONE
38	#21. KAIBAB/TOP LOCALITY	DB1abc	NAVAJO	N65E STRIKE, DIP ~60-70SE & NW	RARE, RAKE ~45E
39	#22. KAIBAB/LONG TRAVERSE	DB1abc	NAVAJO	N55E STRIKE, DIP ~60-80SE & NW	RAKE ~65-90SW
40	#23. KAIBAB/BUTLER VALLEY DRAW	DB1abc	NAVAJO	STRIKE N50E, DIP ~30-65NW	RAKE ~70-85SW
41		DB1abc	NAVAJO	STRIKE N50E, DIP ~60-90SE	RAKE ~70-85SW
42	#24. KAIBAB/ANGELA'S PLACE	DB1c	UPPERMOST NAVAJO	STRIKE N85W, DIP ~30-70N	RARE
43		DB1c	UPPERMOST NAVAJO	STRIKE N60E, DIP ~70-85NW	RARE
44		DB2a	UPPERMOST NAVAJO	BALL OF STRING, DIP ~50-70	NONE
45	#25. KAIBAB/DISCOVERY RIDGE	DB2a	UPPERMOST NAVAJO	BALL OF STRING, DIP ~40-75	NONE
46		DB1a	UPPERMOST NAVAJO	N25E STRIKE, DIP ~70-80	RAKE ~70-90
47	#26. KAIBAB/HACKBERRY CANYON	DB1abc	NAVAJO	N85E STRIKE, DIP ~70-85N	NONE
48	#27. KAIBAB/COYOTE ROAD	DB1abc	STRAIGHT CLIFFS	N75E STRIKE, DIP ~70-85NW & SE	RAKE ~5-15W
49		DB1d	STRAIGHT CLIFFS	N20E STRIKE, DIP ~60-80E	NONE
50		DB1abc	STRAIGHT CLIFFS	N60W STRIKE, DIP ~60-85NE	RAKE ~10-25NW
51	#28. KAIBAB/STRANGELOVE	DB1a	WAHWEAP	N25E-N15W STRIKE, DIP ~50-85W	NONE
52	#29. KAIBAB/WEST SAND COVE	DB1ac	UPPERMOST NAVAJO	N30W STRIKE, DIP ~70-85E	RAKE ~25SE
53		DB1ac	UPPERMOST NAVAJO	N85E STRIKE, DIP ~20-70NW	NONE
54	#30. KAIBAB/POWER LINE	DB1d	NAVAJO	N15E STRIKE, DIP ~45-75SE	NONE
55		DB1abc	NAVAJO	N70E STRIKE, DIP ~75-85NW & SE	RAKE ~5-10SW
56		DB1abc	NAVAJO	N55W STRIKE, DIP ~60-85NE	RAKE ~10-30NW
57					
58					
59	#31. KAIPAROWITS/ALVEY SYNCLINE	DB2b	STRAIGHT CLIFFS	N5W STRIKE, DIP ~60-70E & W	NONE
60	#32. KAIPAROWITS/HELL'S BACKBONE	DB1a	NAVAJO	N35E STRIKE, DIP ~60-85NW	NONE
61	#33. KAIPAROWITS/ESCALANTE MONOCLIN	DB2a	UPPERMOST NAVAJO	BALL OF STRING, DIP ~30-80	NONE
62		DB1a	UPPERMOST NAVAJO	NS STRIKE, DIP ~60-80E	RAKE ~80-90
63	#34. KAIPAROWITS/SEEP FLAT	DB2a	UPPERMOST NAVAJO	BALL OF STRING, DIP ~30-65	NONE
64	#35. KAIPAROWITS/DRY FORK	DB2a	UPPERMOST NAVAJO	N60E STRIKE, DIP ~60-70NW & SE	NONE
65	#36. KAIPAROWITS/BENCHMARK	DB2d	ENTRADA	EW STRIKE, DIP ~45-80S	RARE
66	#37. KAIPAROWITS/DANCE HALL	DB2c	ENTRADA	EW STRIKE, DIP ~40-65S & N	RAKE ~75-90
67		DB2c	ENTRADA	N10W STRIKE, DIP ~60-70E & W	RAKE ~30-45N
68	#38. KAIPAROWITS/FORTYMILE SPRING	DB2c	ENTRADA	N10W STRIKE, DIP ~45-85E	RARE
69	#39. KAIPAROWITS/SOONER ROCKS	DB2cd	ENTRADA	N60W STRIKE, DIP ~45-65NE & SW	RARE, RAKE ~80-90
70	#40. KAIPAROWITS/CAVE SPRINGS	DB2d	ENTRADA	BALL OF STRING, DIP ~30-85	NONE
71	#41. KAIPAROWITS/SODA SPRINGS	DB2d	ENTRADA	BALL OF STRING, DIP ~10-80	NONE
72	#42. KAIPAROWITS/ROCK ROAD	DB2c	UPPERMOST NAVAJO	N20W STRIKE, DIP ~75-85W	NONE
73	#43. KAIPAROWITS/HOLE-IN-THE-ROCK	DB1ab	NAVAJO	N55W STRIKE, DIP ~60-85NE & SW	RAKE ~70-80SE

APPENDIX TABLE 2. (continued)

	F	G	H	I	J
1	KINEMATICS	~SIGMA 1	~SIGMA 2	~SIGMA 3	CAUSE OF DEFORMATION
2	VERT COMPACTION, NORMAL SHEAR, NO HOR EXTENSION	VERTICAL	NA	NA	JURASSIC BURIAL LOADING
3	VERT COMPACTION, NORMAL SHEAR, NO HOR EXTENSION	VERTICAL	NA	NA	JURASSIC DEPOSITION/BURIAL LOADING
4	VERT COMPACTION, NORMAL SHEAR, NO HOR EXTENSION	VERTICAL	NA	NA	JURASSIC DEPOSITIONAL/BURIAL LOADING
5	VERT COMPACTION, NORMAL SHEAR, NO HOR EXTENSION	VERTICAL	NA	NA	JURASSIC DEPOSITIONAL/BURIAL LOADING
6	VERT COMPACTION, NORMAL SHEAR, NO HOR EXTENSION	VERTICAL	NA	NA	JURASSIC DEPOSITIONALBURIAL LOADING
7					
8					
9	VERT COMPACTION, NORMAL SHEAR, NO HOR EXTENSION	VERTICAL	NA	NA	JURASSIC DEPOSITIONAL/BURIAL LOADING
10	VERT COMPACTION, NORMAL SHEAR, NO HOR EXTENSION	VERTICAL	NA	NA	JURASSIC DEPOSITIONAL/BURIAL LOADING
11	VERT COMPACTION, NORMAL SHEAR, NO HOR EXTENSION	VERTICAL	NA	NA	JURASSIC DEPOSITIONAL/BURIAL LOADING
12	VERT COMPACTION, NORMAL SHEAR, N-S EXTENSION (?)	VERTICAL	NA	NA	JURASSIC DEPOSITIONAL/BURIAL LOADING
13	VERT COMPACTION, NORMAL SHEAR, N-S EXTENSION (?)	VERTICAL	NA	NA	JURASSIC DEPOSITIONAL/BURIAL LOADING
14	VERT COMPACTION, NORMAL SHEAR, N55W EXTENSION (?)	VERTICAL	NA	NA	JURASSIC DEPOSITIONAL/BURIAL LOADING
15	VERT COMPACTION, NORMAL SHEAR, N55W EXTENSION (?)	VERTICAL	NA	NA	JURASSIC DEPOSITIONAL/BURIAL LOADING
16	HOR EXTENSION, NORMAL SHEAR	VERTICAL	HOR/N35E	HOR/N55W	BASIN AND RANGE NORMAL FAULTING
17	HOR EXTENSION, NORMAL SHEAR	VERTICAL	HOR/N35E	HOR/N55W	BASIN AND RANGE NORMAL FAULTING
18	VERT COMPACTION, NORMAL SHEAR, NO HOR EXTENSION	VERTICAL	NA	NA	JURASSIC DEPOSITIONAL/BURIAL LOADING
19	HOR EXTENSION, NORMAL SHEAR	VERTICAL	HOR/N35E	HOR/N55W	BASIN AND RANGE NORMAL FAULTING
20	HOR EXTENSION, NORMAL SHEAR	VERTICAL	HOR/N35E	HOR/N55W	BASIN AND RANGE NORMAL FAULTING
21	VERT COMPACTION, NORMAL SHEAR, NO HOR EXTENSION	VERTICAL	NA	NA	JURASSIC DEPOSITIONAL/BURIAL LOADING
22	HOR EXTENSION, NORMAL SHEAR	VERTICAL	HOR/N25E	HOR/N65W	BASIN AND RANGE NORMAL FAULTING
23	HOR EXTENSION, NORMAL SHEAR	VERTICAL	HOR/N25E	HOR/N65W	BASIN AND RANGE NORMAL FAULTING
24					
25					
26	HOR SHORTENING, THRUST SHEAR	HOR/N20E	HOR/N70W	VERTICAL	~20MA THRUST FAULTING RE/TO VOLCANIC COLLAPSE & SPREADING
27	HOR SHORTENING, THRUST SHEAR	HOR/N40W	HOR/N50E	VERTICAL	~20 MA THRUST FAULTING RE/ TO VOLCANIC COLLAPSE & SPREADIN
28	HOR SHORTENING, THRUST SHEAR	HOR/N40W	HOR/N50E	VERTICAL	~20 MA THRUST FAULTING RE/ TO VOLCANIC COLLAPSE & SPREADIN
29	HOR EXTENSION, NORMAL SHEAR	VERTICAL	HOR/N10E	HOR/N80W	BASIN AND RANGE NORMAL FAULTING
30					
31					
32	VERT COMPACTION, NORMAL SHEAR, NO HOR EXTENSION	18/N75W	50/N40E	32/S	JURASSIC DEPOSITIONAL/BURIAL LOADING
33	THRUST/LEFT-HANDED OBLIQUE SHEAR	10/S75E	60/N	25/S20W	LARAMIDE REVERSE/RIGHT-HANDED TRANSPRESSION
34	THRUST/RIGHT-HANDED OBLIQUE SHEAR	10/S75E	60/N	25/S20W	LARAMIDE REVERSE/RIGHT-HANDED TRANSPRESSION
35	BEDDING-PARALLEL THRUST SHEAR	0/W	0/N	VERTICAL	LARAMIDE OUT-OF-THE-SYNCLINE SHEAR RE/ TO KAIBAB MONOCLINE
36	HOR SHORTENING, THRUST SHEAR	HOR, N80E	HOR, N10W	VERT	LARAMIDE SHORTENING NEAR LOWER HINGE OF MONOCLINE
37	HOR SHORTENING, THRUST SHEAR	HOR, N80E	HOR, N10W	VERT	LARAMIDE SHORTENING NEAR LOWER HINGE OF MONOCLINE
38	HOR EXTENSION, NORMAL (OBLIQUE) SHEAR	80/S55W	10/N55E	HOR/N35W	LARAMIDE FOLDING, OUTER ARC STRETCHING OF MONOCLINE
39	HOR EXTENSION, NORMAL (OBLIQUE) SHEAR	80/S55W	10/N55E	HOR/N35W	LARAMIDE FOLDING, OUTER ARC STRETCHING OF MONOCLINE
40	HOR EXTENSION, NORMAL (OBLIQUE) SHEAR	80/S55W	10/N55E	HOR/N35W	LARAMIDE FOLDING, OUTER ARC STRETCHING OF MONOCLINE
41	HOR EXTENSION, NORMAL (OBLIQUE) SHEAR	80/S55W	10/N55E	HOR/N35W	LARAMIDE FOLDING, OUTER ARC STRETCHING OF MONOCLINE
42	HOR EXTENSION, NORMAL (OBLIQUE) SHEAR	80/S55W	10/N55E	HOR/N35W	LARAMIDE FOLDING, OUTER ARC STRETCHING OF MONOCLINE
43	HOR EXTENSION, NORMAL (OBLIQUE) SHEAR	80/S55W	10/N55E	HOR/N35W	LARAMIDE FOLDING, OUTER ARC STRETCHING OF MONOCLINE
44	VERT COMPACTION, NORMAL SHEAR, NO HOR EXTENSION	VERTICAL	NA	NA	JURASSIC DEPOSITIONAL/BURIAL LOADING
45	VERT COMPACTION, NORMAL SHEAR, NO HOR EXTENSION	VERT	NA	NA	JURASSIC DEPOSITIONAL/BURIAL LOADING
46	HOR EXTENSION, NORMAL (OBLIQUE) SHEAR	80/S55W	10/N55E	HOR/N35W	LARAMIDE FOLDING, OUTER ARC STRETCHING OF MONOCLINE
47	HOR EXTENSION, NORMAL (OBLIQUE) SHEAR	80/S55W	10/N55E	HOR/N35W	LARAMIDE FOLDING, OUTER ARC STRETCHING OFMONOCLINE
48	THRUST/RIGHT-HANDED (OBLIQUE) SHEAR	40/N78W	50/E	10/S8W	LARAMIDE REVERSE/RIGHT-HANDED TRANSPRESSION
49	BEDDING-PARALLEL THRUST SHEAR	HOR/N70W	HOR/N20E	VERT	LARAMIDE FOLDING, OUT-OF-THE-SYNCLINE SHEAR ON MONOCLINE
50	THRUST/LEFT-HANDED OBLIQUE SHEAR	40/N78W	50/E	VERT	LARAMIDE REVERSE/RIGHT-HANDED TRANSPRESSION
51	LAYER-PARALLEL SHORTENING VIA FLATTENING	HOR/N80W	0/N10E	VERT	LARAMIDE SHORTENING NEAR LOWER HINGE OF MONOCLINE
52	REVERSE/LEFT-HANDED OBLIQUE SHEAR	5/W	56/N10W	26/S28W	LARAMIDE REVERSE/RIGHT-HANDED TRANSPRESSION
53	REVERSE/RIGHT-HANDED OBLIQUE SHEAR	5/W	56/N10W	26/S28W	LARAMIDE REVERSE/RIGHT-HANDED TRANSPRESSION
54	BEDDING-PARALLEL THRUST SHEAR	HOR/N75W	HOR/N15E	VERT	LARAMIDE FOLDING, OUT-OF-THE-SYNCLINE SHEAR ON MONOCLINE
55	REVERSE/RIGHT-HANDED OBLIQUE SHEAR	45/N82W	42/S88E	8/S2W	LARAMIDE REVERSE/RIGHT-HANDED TRANSPRESSION
56	REVERSE/LEFT-HANDED OBLIQUE SHEAR	45/N82W	42/S88E	8/S2W	LARAMIDE REVERSE/RIGHT-HANDED TRANSPRESSION
57					
58					
59	HOR EXTENSION, NORMAL SHEAR	VERT	0/N5W	0/N85E	LARAMIDE FOLDING, OUTER-ARC EXTENSION
60	HOR EXTENSION, NORMAL SHEAR	VERT	0/N35E	0/N55W	BASIN AND RANGE NORMAL FAULTING
61	VERT COMPACTION, NORMAL SHEAR, NO HOR EXTENSION	VERT	NA	NA	JURASSIC DEPOSITIONAL/BURIAL LOADING
62	HOR CONTRACTION, REVERSE SHEAR	30/E	0/N	60/W	LARAMIDE CONTRACTION ALONG LOWER HINGE OF MONOCLINE
63	VERT COMPACTION, NORMAL SHEAR, NO HOR EXTENSION	VERT	NA	NA	JURASSIC DEPOSITION/BURIAL LOADING
64	HOR EXTENSION, NORMAL SHEAR	VERT	0/N60E	0/N30W	JURASSIC DEPOSITIONAL/BURIAL LOADING
65	HOR EXTENSION, NORMAL SHEAR	VERT	0/E	0/N	LARAMIDE EXTENSION ACCOMPANYING REGIONAL SHORTENING
66	HOR EXTENSION, NORMAL SHEAR	VERT	0/E	0/N	LARAMIDE EXTENSION ACCOMPANYING REGIONAL SHORTENING
67	HOR EXTENSION, NORMAL SHEAR	VERT	0/N10W	0/N80E	LARAMIDE EXTENSION ACCOMPANYING REGIONAL SHORTENING
68	HOR EXTENSION, NORMAL SHEAR	VERT	0/N10W	0/N80E	LARAMIDE EXTENSION ACCOMPANYING REGIONAL SHORTENING
69	HOR EXTENSION, NORMAL SHEAR	VERT	0/N60W	0/N30E	LARAMIDE EXTENSION ACCOMPANYING REGIONAL SHORTENING
70	HOR EXTENSION, NORMAL SHEAR	VERT	?	?	LARAMIDE EXTENSION ACCOMPANYING REGIONAL SHORTENING
71	HOR EXTENSION, NORMAL SHEAR	VERT	?	?	LARAMIDE EXTENSION ACCOMPANYING REGIONAL SHORTENING
72	HOR EXTENSION, NORMAL SHEAR	VERT	0/N20W	0/N70E	JURASSIC DEPOSITIONAL/BURIAL LOADING
73	HOR EXTENSION, NORMAL SHEAR	VERT	0/N55W	0/N35E	LARAMIDE EXTENSION ACCOMPANYING REGIONAL SHORTENING

APPENDIX TABLE 2. (continued)

	A	B	C	D	E
74					
75					
76	#44. WATERPOCKET/ANNIE'S CANYON	DB1ab	NAVAJO	N70E STRIKE, DIP ~70-85SE & NW	RARE, RAKE ~65W
77	#45. WATERPOCKET/BULLFROG	DB2c	ENTRADA	N45W STRIKE, DIP ~50-80NE & SW	RAKE ~70-90
78		DB2c	ENTRADA	N30E STRIKE, DIP ~60SW & NW	RAKE ~85-90
79	#46. WATERPOCKET/AMY'S	DB2c	ENTRADA	N45E STRIKE, DIP ~65-85SE & NW	RARE
80	#47. WATERPOCKET/BRIMHALL BRIDGE	DB2c	ENTRADA	N20W STRIKE, DIP ~35-60NE & SW	RAKE ~70-80
81		DB2c	ENTRADA	N75W STRIKE, DIP ~70-85NE & SW	RARE
82		DB1ab	NAVAJO	N25W STRIKE, DIP ~70-90NE & SW	NONE
83		DB1ab	NAVAJO	N70E STRIKE, DIP ~80NW & SE	RARE
84	#48. WATERPOCKET/THE POST	DB2c	ENTRADA	N80W STRIKE, DIP ~35-70SW	RAKE ~10-30W
85	#49. WATERPOCKET/SURPRISE CANYON	DB2a	UPPERMOST NAVAJO	BALL OF STRING, DIP ~30-80	RARE
86	#50. WATERPOCKET/BURR TRAIL	DB1abcd	NAVAJO	N80W STRIKE, DIP ~55-75S	RAKE ~20-25W
87		DB1abcd	NAVAJO	N55E STRIKE, DIP ~30-65SE	RAKE ~10-20NE
88		DB1d	NAVAJO	N40W STRIKE, DIP~36NE	RARE
89	#51. WATERPOCKET/PEEK-A-BOO	DB1a	NAVAJO	N25W STRIKE, DIP ~45-60NE & SW	RARE
90	#52. WATERPOCKET/LIZARD CANYON	DB2c	ENTRADA	N25W STRIKE, DIP ~40-65NE & SW	RAKE ~85
91	#53. WATERPOCKET/DOGWATER	DB2c	ENTRADA	N75W STRIKE, DIP ~55-65NE & SW	RARE
92	#54. WATERPOCKET/SHEETS GULCH	DB1abc	NAVAJO	N55E STRIKE, DIP ~75NW	RAKE ~0-10
93		DB1abc	NAVAJO	N70W STRIKE, DIP ~70NW	RAKE ~0-10
94	#55. WATERPOCKET/FIVEMILE WASH	DB2c	ENTRADA	EW STRIKE, DIP ~65-75N & S	NONE
95		DB2c	ENTRADA	N40E STRIKE, DIP ~75-85NW & SE	RAKE ~75
96	#56. WATERPOCKET/NOTUM	DB2c	ENTRADA	N40E STRIKE, DIP ~ 60-80SE & NW	NONE
97	#57. WATERPOCKET/PARADISE CREEK	DB2a	UPPERMOST NAVAJO	N55E STRIKE, DIP ~80-85SE	NONE
98		DB2a	UPPERMOST NAVAJO	N15W STRIKE, DIP ~60-80NE & SW	NONE
99	#58. WATERPOCKET/CAPITOL WASH	DB1ac	NAVAJO	N65E STRIKE, DIP ~65-85NW & SE	RAKE ~5-10
100		DB1ac	NAVAJO	N25W STRIKE, DIP ~75-85NE & SW	NONE
101					
102					
103	#59. SAN RAFAEL/BELL CANYON	DB1ab	NAVAJO	N30W STRIKE, DIP ~70-80NE	RAKE ~10-20N
104		DB1ab	NAVAJO	N75W STRIKE, DIP ~70-80NE & SW	RAKE ~10-20W
105		DB2b	NAVAJO	N45E STRIKE, DIP ~60-85 NW & SE	NONE
106	#60. SAN RAFAEL/HORSE CREEK	DB1ab	ENTRADA & NAVAJO	N50W STRIKE, DIP ~55-75NE & SW	RAKE ~80-90
107	#61. SAN RAFAEL/GILSON BUTTE	DB1a	ENTRADA	N50W STRIKE, DIP ~55-75NE & SW	RAKE ~85-90
108	#62. SAN RAFAEL/TEMPLE WASH	DB1ac	NAVAJO	N70W STRIKE, DIP ~50-80NE & SW	RAKE ~20-30W
109	#63. SAN RAFAEL/CAIRN CANYON	DB1a	NAVAJO	N20E STRIKE, DIP ~30W	RAKE ~80-90
110	#64. SAN RAFAEL/NORTH FAULT	DB1ab	NAVAJO	N70E STRIKE, DIP ~60-70SE & NW	RAKE ~70-80
111		DB1ab	NAVAJO	N20W STRIKE, DIP ~60-80NE & SW	RAKE ~70-85
112					
113					
114	#65. HENRY MTNS/HANSEN CREEK	DB2d	ENTRADA	N70W STRIKE, DIP ~45-80SW	RAKE ~60-75W
115	#66. HENRY MTNS/TICABOO	DB2c	ENTRADA	N50E STRIKE, DIP ~70-85NW	RAKE ~55-90W
116	#67. HENRY MTNS/SE ELLSWORTH	DB2c	ENTRADA	N80E STRIKE, DIP ~70-85NW	NONE
117		DB2C	ENTRADA	N40W STRIKE, DIP ~60-85NE & SW	RAKE ~80-90
118	#68. HENRY MTNS/SW ELLSWORTH	DB1a & DB2c	ENTRADA	N65E STRIKE, DIP ~55-75NE & SW	RAKE ~30-60E
119	#69. HENRY MTNS/NW ELLSWORTH	DB1a & DB2c	ENTRADA	N25W STRIKE, DIP ~60-80NE & SW	RARE
120	#70. HENRY MTNS/SW HOLMES	DB1a	ENTRADA	N80E STRIKE, DIP ~55-85SE	RARE
121	#71. HENRY MTNS/WNW HOLMES	DB1ac	ENTRADA	N60W STRIKE, DIP ~60-85NE	RARE
122	#72. HENRY MTNS/N HOLMES	DB1abc	NAVAJO	N10E STRIKE, DIP ~60-85E & W	RAKE ~10-30S
123	#73. HENRY MTNS/LOWER CANYON TRAIL	DB2c	ENTRADA	NS STRIKE, DIP ~60-75E	NONE
124		DB2c	ENTRADA	N50W STRIKE, DIP ~55-80NE	NONE
125	#74. HENRY MTNS/MAIDENWATER RAVINE	DB1acd	NAVAJO	N15W STRIKE, DIP ~25-35E & W	NONE
126	#75. HENRY MTNS/TRAIL CANYON	DB1a	NAVAJO	N75E STRIKE, DIP ~30-80NW	NONE
127	#76. HENRY MTNS/GHOST RIDGE	DB1a & DB2c	ENTRADA	N50W STRIKE, DIP ~65-85NE & SW	VARIABLE
128	#77. HENRY MTNS/COPPER CREEK	DB1a & DB2d	NAVAJO & ENTRADA	ROUGHLY N10E STRIKE, VARIABLE DIF	RARE
129	#78 HENRY MTNS/HILLERS	DB1a & DB2d	NAVAJO & ENTRADA	BALL OF STRING, DIP ~30-85	RAKE ~10-20NE

APPENDIX TABLE 2. (continued)

	F	G	H	I	J
74					
75					
76	HOR EXTENSION, NORMAL SHEAR	VERT	0/N70E	0/N20W	LARAMIDE EXTENSION ACCOMPANYING FORMATION OF MONOCLINE
77	HOR EXTENSION, NORMAL SHEAR	VERT	0/N45W	0/N45W	LARAMIDE EXTENSION ACCOMPANYING REGIONAL SHORTENING
78	HOR EXTENSION, NORMAL SHEAR	VERT	0/N30E	0/N60W	LARAMIDE EXTENSION ACCOMPANYING REGIONAL SHORTENING
79	HOR EXTENSION, NORMAL SHEAR	VERT	0/N45E	0/N45W	LARAMIDE EXTENSION ACCOMPANYING REGIONAL SHORTENING
80	NORMAL AND MINOR REVERSE SHEAR	VERT	0/N20W	0/N70E	LARAMIDE EXTENSION ACCOMPANYING REGIONAL SHORTENING
81	HOR EXTENSION, NORMAL SHEAR	VERT	0/N75W	0/N15E	LARAMIDE EXTENSION ACCOMPANYING REGIONAL SHORTENING
82	HOR EXTENSION, NORMAL SHEAR	VERT	0/N2W	0/N65E	LARAMIDE EXTENSION ACCOMPANYING REGIONAL SHORTENING
83	HOR EXTENSION, NORMAL SHEAR	VERT	0/N75E	0/N15W	LARAMIDE EXTENSION ACCOMPANYING REGIONAL SHORTENING
84	NORMAL AND MINOR REVERSE SHEAR	VERT	0/N80W	0/N10E	LARAMIDE EXTENSION ACCOMPANYING REGIONAL SHORTENING
85	NORMAL AND MINOR REVERSE SHEAR	VERT	NA	NA	JURASSIC DEPOSITIONAL/BURIAL LOADING
86	HOR CONTRACTION, REVERSE, LEFT-HANDED SHEAR	20/S68W	48/S48E	40/N10W	LARAMIDE COMPRESSION DURING FORMATION OF MONOCLINE
87	HOR CONTRACTION, REVERSE, RIGHT-HANDED SHEAR	20/S68W	48/S48E	40/N10W	LARAMIDE COMPRESSION DURING FORMATION OF MONOCLINE
88	THRUST SHEAR	0/N50E	0/N40W	VERT	LARAMIDE SHORTENING, OUT OF SYNCLINE SHEAR
89	NORMAL AND MINOR THRUST SHEAR	VERT	0/N25W	0/N65E	LARAMIDE EXTENSION ACCOMPANYING REGIONAL SHORTENING
90	NORMAL AND MINOR THRUST SHEAR	VERT	0/N25W	0/N65E	LARAMIDE EXTENSION ACCOMPANYING REGIONAL SHORTENING
91	HOR EXTENSION, NORMAL SHEAR	VERT	0/N75W	0/N15E	LARAMIDE EXTENSION ACCOMPANYING REGIONAL SHORTENING
92	HOR CONTRACTION, RIGHT-HANDED SHEAR	0/N80E	VERT	0/N10W	LARAMIDE SHORTENING DURING FORMATION OF MONOCLINE
93	HOR CONTRACTION, LEFT-HANDED SHEAR	0/N80E	VERT	0/N10W	LARAMIDE SHORTENING DURING FORMATION OF MONOCLINE
94	HOR EXTENSION, NORMAL SHEAR	VERT	0/E	0/N	LARAMIDE EXTENSION ACCOMPANYING REGIONAL SHORTENING
95	HOR EXTENSION, NORMAL SHEAR	VERT	0/N40E	0/N50W	LARAMIDE EXTENSION ACCOMPANYING REGIONAL SHORTENING
96	HOR EXTENSION, NORMAL SHEAR	VERT	0/N40E	0/N50W	LARAMIDE EXTENSION ACCOMPANYING REGIONAL SHORTENING
97	VERT COMPACTION, NORMAL SHEAR, NO HOR EXTENSION	VERT	0/N55E	0/N35W	JURASSIC DEPOSITIONAL/BURIAL LOADING
98	VERT COMPACTION, NORMAL SHEAR, NO HOR EXTENSION	VERT	0/N15W	0/N75E	JURASSIC DEPOSITIONAL/BURIAL LOADING
99	HOR EXTENSION, NORMAL SHEAR	VERT	0/N65E	0/N25W	LARAMIDE EXTENSION ASSOCIATED WITH MONOCLINAL FOLDING
100	HOR EXTENSION, NORMAL SHEAR	VERT	0/N35W	0/N65E	LARAMIDE EXTENSION ASSOCIATED WITH MONOCLINAL FOLDING
101					
102					
103	HOR CONTRACTION, LEFT-HANDED SHEAR	0/N50W	VERT	0/N40E	LARAMIDE SHORTENING DURING FORMATION OF MONOCLINE
104	HOR CONTRACTION, RIGHT-HANDED SHEAR	0/N50W	VERT	0/N40E	LARAMIDE SHORTENING DURING FORMATION OF MONOCLINE
105	HOR EXTENSION, NORMAL SHEAR	VERT	0/N45E	0/N45W	JURASSIC DEPOSITIONAL/BURIAL LOADING
106	HOR EXTENSION, NORMAL SHEAR	VERT	0/N50W	0/N40E	LARAMIDE EXTENSION ACCOMPANYING SHORTENING ON UPLIFT
107	HOR EXTENSION, NORMAL SHEAR	VERT	0/N50W	0/N40E	LARAMIDE EXTENSION ACCOMPANYING SHORTENING ON UPLIFT
108	HOR CONTRACTION, RIGHT-HANDED SHEAR	0/N45W	0/N45E	VERT	LARAMIDE SHORTENING DURING FORMATION OF MONOCLINE
109	HOR CONTRACTION, THRUST SHEAR	0/N70W	0/N20E	VERT	LARAMIDE SHORTENING IN LIMB OF MONOCLINE
110	HOR EXTENSION, NORMAL SHEAR	VERT	0/N70E	0/N20W	LARAMIDE EXTENSION ACCOMPANYING FM OF UPLIFT
111	HOR EXTENSION, NORMAL SHEAR	VERT	0/N70E	0/N20W	LARAMIDE EXTENSION ACCOMPANYING FM OF UPLIFT
112					
113					
114	HOR EXTENSION, NORMAL SHEAR	VERT	0/N70W	0/N20E	LARAMIDE EXTENSION ACCOMPANYING REGIONAL SHORTENING
115	HOR EXTENSION, NORMAL SHEAR	VERT	0/N55E	0/N35W	EARLY MIOCENE EXTENSION DURING EMPLACEMENT OF LACCOLITH
116	HOR EXTENSION, NORMAL SHEAR	VERT	0/N80E	0/N10W	LARAMIDE EXTENSION ACCOMPANYING REGIONAL SHORTENING
117	HOR EXTENSION, NORMAL SHEAR	VERT	0/N40W	0/N50E	EARLY MIOCENE EXTENSION DURING EMPLACEMENT OF LACCOLITH
118	HOR EXTENSION, NORMAL SHEAR	VERT	0N65E	0/N25W	EARLY MIOCENE EXTENSION DURING EMPLACEMENT OF LACCOLITH
119	HOR EXTENSION, NORMAL SHEAR	VERT	0/N25W	0/N65E	EARLY MIOCENE EXTENSION DURING EMPLACEMENT OF LACCOLITH
120	HOR EXTENSION, NORMAL SHEAR	VERT	0/N80E	0/N10W	EARLY MIOCENE EXTENSION DURING EMPLACEMENT OF LACCOLITH
121	HOR EXTENSION, NORMAL SHEAR	VERT	0/N60W	0/N30E	EARLY MIOCENE EXTENSION DURING EMPLACEMENT OF LACCOLITH
122	HOR EXTENSION, NORMAL SHEAR	VERT	0/N10E	0/N80W	EARLY MIOCENE EXTENSION DURING EMPLACEMENT OF LACCOLITH
123	HOR EXTENSION, NORMAL SHEAR	VERT	0/N	0/E	EARLY MIOCENE EXTENSION DURING EMPLACEMENT OF LACCOLITH (?)
124	HOR EXTENSION, NORMAL SHEAR	VERT	0/N50W	0/N40E	EARLY MIOCENE EXTENSION DURING EMPLACEMENT OF LACCOLITH (?)
125	HOR CONTRACTION, THRUST SHEAR	0/N75E	0/N15W	0/N75E	EARLY MIOCENE SHORTENING NEAR END OF MAGMA TONGUE
126	HOR EXTENSION, NORMAL SHEAR	VERT	0/N75E	0/N15W	EARLY MIOCENE EXTENSION DURING EMPLACEMENT OF LACCOLITH
127	HOR EXTENSION, NORMAL SHEAR	VERT	0/N50W	0/N40E	EARLY MIOCENE EXTENSION DURING EMPLACEMENT OF LACCOLITH
128	HOR EXTENSION, NORMAL SHEAR	VERT	0/N10E	0/N80W	EARLY MIOCENE EXTENSION DURING EMPLACEMENT OF LACCOLITH
129	HOR EXTENSION, NORMAL SHEAR	VERT	?	?	EARLY MIOCENE EXTENSION DURING EMPLACEMENT OF LACCOLITH

ACKNOWLEDGMENTS

The field work that provides the foundation for descriptions and interpretations presented in this Geological Society of America Special Paper was carried out during the summers of 1985–1990 and more intensively during the summers of 1994–1997. The research was supported primarily by three sources: the Petroleum Research Fund of the American Chemical Society, the Earth Sciences Division of the National Science Foundation, and the Department of Geosciences at the University of Arizona. I express appreciation to the donors of the Petroleum Research Fund, administered by the American Chemical Society, for two grants: "Thin-Skinned Deformation in the Bryce-Zion Region of Southern Utah" (1998–1991) and "Characterization of Deformation-Band Controlled Connectivity and Compartmentalization in Sandstone" (1996–1998). In addition, I express my appreciation to the National Science Foundation for support of this work through NSF #EAR-9406208, "Use of Deformation Bands as a Guide to Regional Tectonic Stress Patterns Within the Southwestern Colorado Plateau" (1994–1997). And third, I express my appreciation to the Department of Geosciences of the University of Arizona, and especially Joaquin Ruiz, Chair, for supporting additional costs of field work and preparation of illustrations.

A part of my field research in the 1980s included collaboration with Richard Nickelsen and Olivier Merle. Their work advanced understanding of Miocene thrusting in the Paunsaugunt Plateau region in important ways. In addition, Robert Krantz and Steve Naruk provided assistance to me in reconnaissance efforts. Eric Lundin carried out significant and important mapping of the Rubys Inn thrust, and he discovered some magnificent localities where the thrust relationships are well exposed. During my field research in the 1990s, I was assisted enormously by students, including high school students, undergraduate students, and graduate students. Those who accompanied me in the field and helped in the collection of data and in mapping are Angela Smith, Kerry Caruthers, Justin Banz, David Dan Davis, Sarah Tindall, Jessica Graybill, Seth Gering, Shari Christofferson, Danielle Vanderhorst, Pilar García, Alex Bump, Scott Witthoft, Karen Swanberg, and Steve Ahlgren. Mark Marekos set up a FoxPro data management system, and for this contribution I am greatly indebted. Jennifer Swenson ran the paleo-focal mechanism solutions that are featured in the final figure of this paper. Furthermore, I also thank my friends Tim Tilton (political scientist) and Ralph Colwell (musician) for their participation in some of the field work; they always wanted to know what it is that I do "out in the field."

I call special attention to Sarah Tindall, who worked with me for two summer field seasons, significantly advancing the project, and who as a research assistant was responsible for substantial data input and compilation. I thank as well Pilar García, who as a research assistant carried out photogeologic mapping of deformation-band shear zones and contributed importantly to structural mapping on the ground.

Through the Cooper Aerial Survey Company (Tucson, Arizona) I contracted flyovers of two special study areas, the Cottonwood area and the Sheets Gulch area. Cooper Aerial Survey Company then prepared outstanding large-scale rectified aerial photographs and topographic maps for the subareas. Jim Holmlund of Geo-Map, Inc. (Tucson, Arizona) was of enormous help in the planning of the flyovers and in personally carrying out GPS surveying of targets in very steep terrane. Susie Gillatt of Terra Chroma, Inc. carried out much of the illustration preparation, including scanning and touch-up of slides and prints, scanning of line drawings prepared in past years by Wally Varner, and drafting of some new line drawings. I appreciate her immense talent and the care and effectiveness with which she carried out this work. Peter J. Kresan did the effective close-up photography of hand specimens.

Special thanks go to friend and colleague Jim Holmlund of Geo-Map, Inc. (Tucson, Arizona) for his skill, care, and artistic sense in preparing the digital cartographic presentations that represent a core element of this paper. He went above and beyond what anyone could normally expect, taking a deep interest in the project.

I acknowledge the strong support received from William G. Higgs, Geology Division, Chevron Petroleum Technology Company, and Charles F. Kluth, Structural Geology School, Chevron Oversees Petroleum, Inc. Chuck and Bill were conveners of the American Association of Petroleum Geologists' Hedberg Research Conference on "Reservoir-Scale Deformation—Characterization and Prediction," which was held at Bryce Canyon in 1997. They supported the preparation costs of some of the illustrations for a field-trip guidebook that I prepared, and many of these illustrations are included within this Geological Society of America Special Paper. Furthermore, the Geological Society of America contributed to illustration preparation related to the development of a field guidebook for their Penrose Conference on "Tectonics of Continental Interiors." The costs for publishing the five false-color satellite images were covered by Chevron. I am *very* appreciative of this support.

Certain rangers, ranchers, and naturalists were very helpful to me and my students during our field work. I especially thank Susan Colclazer, who was Head of Interpretation, and her staff at Bryce Canyon National Park; Cydny Quitter, Ranger, Kodachrome Basin State Park; Steve Dalton, Manager, the Sandy Ranch; Thomas Clark, Chief of Resource Management and Science, and his staff, Capitol Reef National Park; the Bureau of Land Management Office, Escalante; and the Utah Geological Survey, particularly William Lund.

I received excellent suggestions from reviewers during the completion stage of this project. These suggestions have improved the manuscript significantly, although I alone remain responsible for shortcomings. One reviewer, James Faulds, captured the intentions of this research in such a useful and succinct way that I used his language in the last paragraph of my abstract, and I acknowledge it here. Abhijit Basu, Science Editor for GSA Books, was a pleasure to work with and guided the review in ways that I greatly value. Equally supportive and skilled, Sonia Smith guided the publication process extremely well as managing editor. Mary Eberle brilliantly copyedited the manuscript.

Finally, I want to extend my deep appreciation to my wife, Merrily, from whom I was away during many months of field work. She knows that I have a love affair with the Colorado Plateau and generously gave me her full support throughout the project.

REFERENCES CITED

Anderson, J. J., 1993, The Markagunt Megabreccia: Large Miocene gravity slides mantling the northern Markagunt Plateau, southwestern Utah: Utah Geological Survey Miscellaneous Publication 93-2, 37 p.

Anderson, J. J., Rowley, P. D., Blackman, J. T., Mehnert, H. H., and Grant, T. C., 1990, Geologic map of the Circleville Canyon area, southern Tushar Mountains and northern Markagunt Plateau, Beaver, Garfield, Iron, and Piute Counties, Utah: U.S. Geological Survey Miscellaneous Investigations Map I-2000.

Anderson, L. W., and Miller, D. G., 1979, Age and extent of Quaternary faulting in Utah: Geological Society of America Abstracts with Programs: v. 11, no. 3, p. 66.

Anderson, R. E., and Barnhard, T. P., 1986, Genetic relationship between faults and folds and determination of Laramide and neotectonic paleostress, western Colorado Plateau transition zone: Tectonics, v. 5, no. 2, p. 335–357.

Antonellini, M., and Aydin, A., 1994, Effect of faulting on fluid flow in porous sandstones: Petrophysical properties: American Association of Petroleum Geologists Bulletin, v. 78, p. 355–377.

Antonellini, M., and Aydin, A., 1995, Effect of faulting on fluid flow in porous sandstones: Geometry and spatial distribution: American Association of Petroleum Geologists Bulletin, v. 79, p. 642–671.

Antonellini, M., Aydin, A., and Pollard, D. D., 1994, Microstructure of deformation bands in porous sandstones at Arches National Park: Journal of Structural Geology, v. 16, p. 941–959.

Arabasz, W. V., and Julander, D. R., 1986, Geometry of seismically active faults and crustal deformation within the Basin and Range—Colorado Plateau transition in Utah, *in* Mayer, L., ed., Extensional tectonics of the southwestern United States; a perspective on processes and kinematics: Geological Society of America Special Paper, v. 208, p. 43–74.

Aydin, A., 1978, Small faults formed as deformation bands in sandstone: Pure and Applied Geophysics, v. 116, p. 913–930.

Aydin, A., and Johnson, A. M., 1978, Development of faults as zones of deformation bands and as slip surfaces in sandstone: Pure and Applied Geophysics, v. 116, p. 931–942.

Aydin, A., and Johnson, A. M., 1983, Analysis of faulting in porous sandstones: Journal of Structural Geology, v. 5, p. 19–31.

Babenroth, D. L., and Strahler, A. N., 1945, Geomorphology of the East Kaibab monocline, Arizona and Utah: Geological Society of America Bulletin, v. 56, p. 107–150.

Billingsley, G. H., Huntoon, P. W., and Breed, W. J., 1987, Geologic map of Capitol Reef National Park and vicinity, Emery, Garfield, Millard, and Wayne Counties, Utah: Utah Geological and Mineral Survey, 5 sheets.

Boardman, D. A., 1995, Deformation bands as indicators of stress and strain during monoclinal folding: Northern end of the East Kaibab monocline, Utah (Senior thesis): Northfield, Minnesota, Carleton College, 56 p.

Borg, I. Y., 1960, Experimental deformation of St. Peter sand: A study of cataclastic flow, *in* Griggs, D. T., and Handin, J., eds., Rock deformation—A symposium: Geological Society of America Memoir 79, p. 133–191.

Borgia, A., Ferrari, L., and Pasquara, G. P., 1992, Importance of gravitational spreading in the tectonic and volcanic evolution of Mount Etna: Nature, v. 357, p. 231–235.

Bowers, W. E., 1972, The Canaan Peak, Pine Hollow, and Wasatch Formations in the Table Cliff region, Garfield County, Utah: U.S. Geological Survey Bulletin 1331-B, 39 p.

Bowers, W. E., 1991, Geologic map of Bryce Canyon National Park and vicinity, southwestern Utah: U.S. Geological Survey Miscellaneous Investigations Series Map I-2108, scale 1:24,000.

Bump, A., Ahlgren, S. G., and Davis, G. H., 1997, A tale of two uplifts: Waterpocket Fold, Capitol Reef National Park: Eos (Transactions, American Geophysical Union), v. 78, no. 46 (Fall Meeting), p. F701.

Byerlee, J. D., and Brace, W. F., 1968, Stick slip, stable sliding, and earthquakes—Effects of rock type, pressure, strain rate, and stiffness: Journal of Geophysical Research, v. 73, p. 6031–6037.

Cashion, W. B., 1961, Geology and fuel resources of the Orderville-Glendale area, Kane County, Utah: U.S. Geological Survey Map C-49.

Cashion, W. B., 1967, Geologic map of the southern flank of the Markagunt Plateau, NW Kane County, Utah: U.S. Geological Survey Map I-494.

Christofferson, S., and Davis, G. H., 1997, Antithetic faulting associated with major strike-slip faulting along the East Kaibab monocline, Candyland, Utah: Eos (Transactions, American Geophysical Union), v. 78, no. 46 (Fall Meeting), p. F701–702.

Coney, P. J., 1978, Mesozoic-Cenozoic Cordilleran plate tectonics, *in* Smith, R. B., and Eaton, G. P., eds., Cenozoic tectonics and regional geophysics of the western Cordillera: Geological Society of America Memoir 152, p. 33–50.

Cook, E. F., 1960, Geologic atlas of Utah, Washington County: Utah Geological and Mineralogical Survey Bulletin 70, 119 p.

Davis, G. H., 1978, The monocline fold pattern of the Colorado Plateau, *in* Matthews, V., ed., Laramide folding associated with basement block faulting in the western U. S.: Geological Society of America Memoir 151, p. 215–233.

Davis, G. H., 1990, Shortening of the Claron Formation at Bryce Canyon, Utah: Geological Society of America Abstracts with Programs, v. 22, no. 3, p. 17.

Davis, G. H., 1996a, Deformation bands: Agents of regional macroscopic ductility: Geological Society of America Abstracts with Programs, v. 28, p. 59–60.

Davis, G. H., 1996b, "Riedel relays" in deformation band shear zones, Colorado Plateau, Utah: Geological Society of America Abstracts with Programs, v. 28, no. 7, p. A-188.

Davis, G. H., 1997a, Field guide to geologic structures in the Bryce Canyon region, Utah: American Association of Petroleum Geologists Conference on Reservoir-Scale Deformation: Characterization and Prediction: Tucson, University of Arizona, 119 p.

Davis, G. H., 1997b, Field guide to geologic structures in the Zion–Bryce–Cedar Breaks region, Utah: Geological Society of America Penrose Conference on Tectonics of Continental Interiors, Brian Head, Utah, 126 p.

Davis, G. H., 1998, Fault-fin landscape: Geological Magazine, v. 135, p. 283–286.

Davis, G. H., and Kiven, C. W., 1975, Structure map of folds in Phanerozoic rocks, Colorado Plateau tectonic province of Arizona: Phoenix, Arizona Oil and Gas Conservation Commission, scale 1:500,000.

Davis, G. H., and Krantz, R. W., 1986, Post-"Laramide" thrust faults in the Claron Formation, Bryce Canyon National Park, Utah: Geological Society of America Abstracts with Programs, v. 18, p. 98.

Davis, G. H., and Reynolds, S. J., 1996, Structural geology of rocks and regions (2nd edition): New York, John Wiley and Sons, 776 p.

Davis, G. H., and Rowley, 1993, Miocene thrusting, gravity sliding, and near-surface batholithic emplacement, Marysvale volcanic field, southwestern Utah: Eos (Transactions, American Geophysical Union), v. 74, p. 647.

Davis, G. H., and Tindall, S. E., 1996, Discovery of major right-handed Laramide strike-slip faulting along the eastern margin of the Kaibab uplift, Colorado Plateau, Utah: Eos (Transactions, American Geophysical Union), v. 77, no. 46, p. F641–642.

Davis, G. H., Showalter, S. R., Benson, G. S., McCalmont, L. C., and Cropp, F. W., 1981, Guide to the geology of the Salt River Canyon, Arizona: Arizona Geological Society Digest, v. 13, p. 48–97.

Davis, G. H., García, P., Bump, A., Ahlgren, S., and Swanberg, K., 1997, Detailed internal structure of a conjugate strike-slip deformation band shear zone system in Navajo Sandstone at a salient in the Waterpocket Fold, Capitol Reef National Park, Utah: Eos (Transactions, American Geophysical Union), v. 78, p. F701.

Davis, G. H., Bump, A., García, P., and Ahlgren, S., 1999, Conjugate Riedel shear systems: Journal of Structural Geology (in press).

Davison, I., 1994, Linked fault systems; extensional, strike-slip and contractional, *in* Hancock, P. L., ed., Continental deformation: Oxford, Pergamon Press, p. 121–142.

Dickinson, W. R., 1989, Tectonic setting of Arizona through geologic time, *in* Jenney, J. P., and Reynolds, S. J., eds., Geologic evolution of Arizona: Arizona Geological Digest, v. 17, p. 1–16.

Dunn, D. E., La Fountain, L. W., and Jackson, R. E., 1973, Porosity dependence and mechanism of brittle fracture in sandstones: Journal of Geophysical Research, v. 78, p. 2403–2417.

Dutton, C. E., 1880, Report on geology of the high plateaus: Publication of the Powell Survey, Government Printing Office, Washington, D. C., 307 p.

Edwards, H. E., Becker, A. D., and Howell, J. A., 1993, Compartmentalization of an aeolian sandstone by structural heterogeneities: Permo-Triassic Hopeman Sandstone, Moray Firth, Scotland, *in* North, C. P., and Prosser, D. J., eds., Characterization of fluvial and aeolian reservoirs: Geological Society [London] Special Publication 73, p. 339–365.

Engdahl, E. R., and Rinehart, W. A., 1991, Seismicity map of North America project, *in* Slemmons, D. B., Engdahl, E. R., Zoback, M. D., and Blackwell, D. D., eds., Neotectonics of North America: Geological Society of America, Boulder, p. 21–27.

Engebretson, D. C., Cox, A., and Gordon, R. G., 1985, Relative motions between oceanic and continental plates in the Pacific Basin: Geological Society of America Special Paper 206, 59 p.

Erslev, E. A., 1991, Trishear fault propagation folding: Geology, v. 19, p. 617–620.

Friedman, M., and Logan, J. M., 1973, Lüders' bands in experimentally deformed sandstone and limestone: Geological Society of America Bulletin, v. 84, p. 1465–1476.

Fossen, H., and Hesthammer, J., 1997, Geometric analysis and scaling relations of deformation bands in porous sandstones: Journal of Structural Geology, v. 19, p. 1479–1493.

García, P., and Davis, G. H., 1997, Detailed internal structure of a conjugate normal deformation band shear zone system in Navajo Sandstone at the northern end of the East Kaibab monocline, southern Utah: Eos (Transactions, American Geophysical Union), v. 78, p. F702.

Gilbert, G. K., 1877, Report on the geology of the Henry Mountains: U.S. Geographic and Geological Survey of Rocky Mountains Region, 170 p.

Gilluly, J., 1928, Geology and oil and gas prospects of part of the San Rafael Swell, Utah: U.S. Geological Survey Bulletin 806-C, p. 69–130.

Gregory, H. E., 1950, Geology and geography of the Zion National Park region, Utah and Arizona: U.S. Geological Survey Professional Paper 220, 200 p.

Gregory, H. E., 1951, The geology and geography of the Paunsaugunt region, Utah: U.S. Geological Survey Professional Paper 226, 160 p.

Higgs, W. G., Williams, G. D., and Powell, C. M., 1991, Evidence for flexural shear folding associated with extensional faults: Geological Society of America Bulletin, v. 103, p. 710–717.

Hintze, L. F., 1980, Geologic map of Utah: Salt Lake City, Utah Geological and Mineral Survey, scale 1:500,000.

Hintze, L. F., 1988, Geologic history of Utah: Brigham Young University Geology Studies Special Publication 7, 202 p.

Hintze, L. F., Anderson, R. E., and Embree, G. F., 1994, Geologic map of the Motoqua and Gunlock quadrangles, Washington County, Utah: U.S. Geological Survey Miscellaneous Investigations Series Map I-2427, 1 plate, scale 1:24,000.

Hunt, C. B., 1946, Guidebook to the geology and geography of the Henry Mountains region, Utah: Utah Geological Society Guidebook, no. 1, 51 p.

Hunt, C. B., 1983, Development of the La Sal and other laccolithic mountains on the Colorado Plateau, *in* Averett, W. R., ed., Northern Paradox Basin–Uncompaghre uplift: Grand Junction Geological Society Guidebook, p. 29–32.

Hunt, C. B., 1988a, Geology of the Henry Mountains, Utah, as recorded in the notebooks of G. K. Gilbert, 1875–76: Geological Society of America Memoir 167, 229 p.

Hunt, C. B., 1988b, The laccolith-stock controversy: New results from the southern Henry Mountains, Utah: Discussion and reply: Geological Society of America Bulletin, v. 100, p. 1657–1658.

Hunt, C. B., Averitt, P., and Miller, R. L., 1953, Geology and geography of the Henry Mountains region, Utah: U.S. Geological Survey Professional Paper 228, 234 p.

Huntoon, P. W., 1993, Influence of inherited Precambrian basement structure on the localization and form of Laramide monoclines, Grand Canyon, Arizona, *in* Schmidt, C. J., Chase, R. B., and Erslev E. A., eds., Laramide basement deformation in the Rocky Mountain foreland of the western United States: Geological Society of America Special Paper 280, p. 243–256.

Huntoon, P. W., and Sears, J. W., 1975, Bright Angel and Eminence faults, eastern Grand Canyon, Arizona: Geological Society of America Bulletin, v. 86, p. 465–472.

Hurlow, H. A., 1998, The geology of the Central Virgin River basin, southwestern Utah, and its relation to ground-water conditions: Utah Geological Survey Water-Resources Bulletin 26, 53 p.

Jackson, M. D., 1998, Processes of laccolithic emplacement in the southern Henry Mountains, southeastern Utah, *in* Friedman, J. D., and Huffman, Jr., A. C., eds., Laccolithic complexes of southeastern Utah: Time of emplacement and tectonic setting—Workshop proceedings: U.S. Geological Survey Bulletin 2158, p. 51–59.

Jackson, M. D., and Pollard, D. D., 1988a, The laccolith-stock controversy: New results from the southern Henry Mountains, Utah: Geological Society of America Bulletin, v. 100, p. 117–139.

Jackson, M. D., and Pollard, D. D., 1988b, The laccolith-stock controversy: New results from the southern Henry Mountains, Utah: Discussion and reply: Geological Society of America Bulletin, v. 100, p. 1658–1659.

Jackson, M. D., and Pollard, D. D., 1990, Flexure and faulting of sedimentary host rocks during growth of igneous domes, Henry Mountains, Utah: Journal of Structural Geology, v. 12, p. 185–206.

Jamison, W. R., 1979, Laramide deformation of the Wingate Sandstone, Colorado National Monument: A study of cataclastic flow [Ph.D. thesis]: College Station, Texas A&M University, 170 p.

Jamison, W. R., and Stearns, D. W., 1982, Tectonic deformation of Wingate Sandstone, Colorado National Monument: American Association of Petroleum Geologist Bulletin, v. 66, p. 2584–2608.

Kelley, V. C., and Clinton, N. J., 1960, Fracture systems and tectonic elements of the Colorado Plateau: University of New Mexico Publications in Geology, no. 6, 104 p.

Krantz, R. W., 1986, The odd-axis model: Orthorhombic fault patterns and three-dimensional strain fields [Ph.D. thesis]: Tucson, The University of Arizona, 98 p.

Krantz, R. W., 1988, Multiple fault sets and three-dimensional strain: Theory and application: Journal of Structural Geology, v. 10, p. 225–237.

Lundin, E. R., 1987, Thrusting of the Claron Formation, the Bryce Canyon region, Utah [M.S. thesis]: The University of Arizona, Tucson, 51 p.

Lundin, E. R., 1989, Thrusting of the Claron Formation, the Bryce Canyon region, Utah: Geological Society of America Bulletin, v. 101, p. 1038–1050.

Lundin, E. R., and Davis, G. H., 1987, Southeast vergent thrust faulting and folding of the Eocene (?) Claron Formation, Bryce Canyon National Park region, Utah: Geological Society of America Abstracts with Programs, v. 19, p. 317.

Marzolf, J. E., 1983, Changing wind and hydrologic regimes during deposition of the Navajo and Aztec Sandstones, Jurassic(?), Southwestern United States, *in* Brookfield, M. E., and Ahlbrandt, T. S. (eds), Eolian sediments and processes: 11th International Association of Sedimentologists Congress, v. 38, p. 635–660.

Merle, O. R., Davis, G. H., Nickelsen, R. P., and Gourlay, P. A., 1993, Relation of thin-skinned thrusting of Colorado Plateau strata in southwestern Utah to Cenozoic magmatism: Geological Society of America Bulletin, v. 1050, p. 387–398.

Mollema, P-N., and Antonellini, M., 1996, Compaction bands; a structural analog for anti-mode I cracks in aeolian sandstone: Tectonophysics, v. 267, p. 209–228.

Naeser, C. W., 1978, Fission track dating: U.S. Geological Survey Open-File Report 76-190, 87 p.

Nelson, S. T., 1998, Re-evaluation of the central Colorado Plateau laccoliths in the light of new age determinations, *in* Friedman, J. D., and Huffman, A. C., Jr., eds., Laccolithic complexes of southeastern Utah: Time of emplacement and tectonic setting—Workshop proceedings: U.S. Geological Survey Bulletin 2158, p. 37–39.

Nickelsen, R. P., Merle, O., and Davis, G. H., 1992, Structures of the mid-Tertiary radial compression in the High Plateaus of southern Utah in relation to Sevier, Laramide, and Basin and Range deformation: Geological Society of America Abstracts with Programs, v. 24, p. 55.

Peacock, D. C. P., and Sanderson, D. J., 1994, Geometry and development of relay ramps in normal fault systems: American Association of Petroleum Geologists Bulletin, v. 78, p. 147–165.

Powell, J. W., 1875, Exploration of the Colorado River of the West: Government Printing Office, Washington, D. C., 291 p.

Reches, Z., 1978, Analysis of faulting in three-dimensional strain field: Tectonophysics, v. 47, p. 109–129.

Reches, Z., 1983, Faulting of rocks in three-dimensional strain fields: II. Theoretical analysis: Tectonophysics, v. 95, p. 133–156.

Saleeby, J. B., Busby-Spera, C., Oldow, J. S., Dunne, J. S., Dunne, G. C., Wright, J. E., Cowan, D. S., Walker, N. W., and Allmendinger, R. W., 1992, Early Mesozoic tectonic evolution of the Western U. S. Cordillera, *in* Burchfiel, B. C., Lipman, P. W., and Zoback, M. L., eds., The Cordilleran orogen: Coterminous U. S.: Boulder, Colorado, Geological Society of America, Geology of North America, p. 107–168.

Sargent, K. A., and Hansen, D. E., 1982, Bedrock geologic map of the Kaiparowits coal-basin area, Utah: U.S. Geological Survey Miscellaneous Investigations Map I-1033-I.

Sargent, K. A., and Philpott, B. C., 1987, Geologic map of the Kanab Quadrangle, Kane County, Utah, and Mohave and Coconino Counties, Arizona: U.S. Geological Survey Quadrangle Map GQ-1603.

Severinghaus, J., and Atwater, T., 1990, Cenozoic geometry and thermal state of the subducting slabs beneath western North America, *in* Wernicke, B., ed., Basin and Range extensional tectonics near the latitude of Las Vegas, Nevada: Geological Society of America Memoir 176, p. 1–22.

Smith, R. B., and Arabasz, W. J., 1991, Seismicity of the Intermountain Seismic Belt, *in* Slemmons, D. B., Engdahl, E. R., Zoback, M. D., and Blackwell, D. D. (eds), Neotectonics of North America: Geological Society of America, Boulder, p. 185–228.

Smith, J. F., Lyman, C., Hinrichs, E. N., and Luedke, R. G., 1963, Geology of the Capitol Reef area, Wayne and Garfield Counties, Utah: U.S. Geological Survey Professional Paper 363, 102 p.

Stewart, M. E., and Taylor, W. J., 1996, Structural analysis and fault segment boundary identification along the Hurricane fault in southwestern Utah: Journal of Structural Geology, v. 18, no. 8, p. 1017–1029.

Stock, J., and Molnar, P., 1988, Uncertainties and implications of the Late Cretaceous and Tertiary position of North America relative to the Farallon, Kula, and Pacific plates: Tectonics, v. 7, p. 1339–1384.

Sullivan, K. R., 1998, Isotopic ages of igneous intrusions in southeastern Utah, *in* Friedman, J. D., and Huffman, A. C., Jr., eds., Laccolithic complexes of southeastern Utah: time of emplacement and tectonic setting—Workshop proceedings: U.S. Geological Survey Bulletin 2158, p. 33–39.

Tindall, S. E., and Davis, G. H., 1999, Monocline development by oblique-slip fault-propagation folding: The East Kaibab monocline, Colorado Plateau, Utah: Journal of Structural Geology, v. 21, p. 1303–1320.

Tosdal, R. M., Haxel, G. B., and Wright, J. E., 1989, Jurassic geology of the Sonoran Desert region, southern Arizona, southeastern California, and northernmost Sonora: Construction of a continental margin magmatic arc, *in* Jenney, J. P. and Reynolds, S. J., eds., Geologic evolution of Arizona: Arizona Geological Digest, v. 17, p. 397–434.

Witkind, I. J., 1991, Implications of distinctive fault sets in the San Rafael Swell and adjacent areas, east-central Utah, *in* Chidsey, T. C., Jr., ed., Geology of east-central Utah: Utah Geological Association Publication, p. 141–148.

Witkind, I. J., 1994, The role of salt in the structural development of central Utah: U.S. Geological Survey Professional Paper 1994, 145 p.

Manuscript Accepted by the Society April 27, 1999

Printed in U.S.A.